In Praise of System-Level Design with Rosetta

The use of heterogeneous models is becoming essential in modern system specification and design. With its rich and extensible set of domains, Rosetta provides a flexible yet rigorous formalism for capturing and bringing together behavior, architecture and constraint definitions from a wide spectrum of diverse engineering sciences.

—Dr. Roberto Passerone *University of Trento*

Designing a language is both very difficult and time consuming. Dr. Perry Alexander has shown great commitment and courage in leading the Rosetta Language project to its successful conclusion while commercial EDA companies and some industry analysts had given up on the project.

Whether the language will ever attain the commercial success it deserves is immaterial. Perry's work and that of his two closest collaborators, David Barton and Peter Ashenden, has not only produced a language that allows the rigorous architecture and specification of a system, but its characteristics will influence the syntax and semantics of all future system design languages. Although the case studies presented focus on electronic design, after all the three primary architects are active in the field of Electronic Design Automation, the Rosetta language has a true "system" nature that allows it to be used as a specification language in other domains.

Teaching a new language is difficult: after all one only learns a language when one uses it, and yet how can one use it if one does not know it? System-Level Design with Rosetta is an excellent tool for technologists who wish to become Rosetta speakers.

Perry is too modest in his Acknowledgment when it says that Rosetta "actually resembles a language." It not only is a language, it is better than most."

—Gabe Moretti *Senior Member IEEE and EDA industry expert*

Real system designers know that the requirements placed upon the design of advanced electronic products extends well beyond the bounds of functional behavior typical in most other "system-level" languages. Parametric constraints such as cost, weight, power, timing, compliance to standards, and numerous domain-specific requirements often dominate written specifications, yet these have been left out of the automated process of design refinement tradeoffs in EDA tool flows—until now. Rosetta is a breakthrough technology, unique in it's breadth of expressiveness and adaptability to differing application domains. Firmly rooted in formal semantics, Rosetta promises to enable better scalability and modularity in handling increasingly large and complex electronic systems, whether confined within a single chip or extending through an aircraft carrier. Perry Alexander and his expert team have labored diligently to capture the full needs of system designers and enable representation within a language that is architected for future extensibility.

—Steven E. Schulz *President and CEO of Silicon Integration Initiative, Inc.*

Formal methods for system-level design are becoming increasingly important as the complexity of systems escalates. The development of Rosetta is a significant milestone, as it allows designers to capture complex interactions among multiple domains involved in real-world systems in a way that can be formally analyzed by tools. Alexander's book leads the reader into this exciting new world, explaining both the language and its use in a variety of modeling scenarios. I am pleased to recommend the book to those with an interest in system-level design, whether they be new to the area or experienced practitioners.

—Peter J. Ashenden *Ashenden Designs Pty. Ltd.*

The Morgan Kaufmann Series in Systems on Silicon

Series Editor: Wayne Wolf, *Princeton University*

The rapid growth of silicon technology and the demands of applications are increasingly forcing electronics designers to take a systems-oriented approach to design. This has lead to new challenges in design methodology, design automation, manufacture and test. The main challenges are to enhance designer productivity and to achieve correctness on the first pass. *The Morgan Kaufmann Series in Systems on Silicon* presents high-quality, peer-reviewed books authored by leading experts in the field who are uniquely qualified to address these issues.

The Designer's Guide to VHDL, Second Edition
Peter J. Ashenden

The System Designer's Guide to VHDL-AMS
Peter J. Ashenden, Gregory D. Peterson, and Darrell A. Teegarden

Readings in Hardware/Software Co-Design
Edited by Giovanni De Micheli, Rolf Ernst and Wayne Wolf

Modeling Embedded Systems and SoCs
Axel Jantsch

ASIC and FPGA Verification: A Guide to Component Modeling
Richard Munden

Multiprocessor Systems-on-Chips
Edited by Ahmed Amine Jerraya and Wayne Wolf

Comprehensive Functional Verification
Bruce Wile, John Goss, and Wolfgang Roesner

Customizable Embedded Processors: Design Technologies and Applications
Edited by Paolo Ienne and Rainer Leupers

Networks on Chips: Technology and Tools
Giovanni De Micheli and Luca Benini

Designing SOCs with Configured Cores: Unleashing the Tensilica Diamond Cores
Steve Leibson

VLSI Test Principles and Architectures: Design for Testability
Edited by Laung-Terng Wang, Cheng-Wen Wu, and Xiaoqing Wen

Contact Information

Charles B. Glaser
Senior Acquisitions Editor
Elsevier
(Morgan Kaufmann; Academic Press; Newnes)
(781) 313-4732
c.glaser@elsevier.com
http://www.books.elsevier.com

Wayne Wolf
Professor
Electrical Engineering, Princeton University
(609) 258 1424
wolf@princeton.edu
http://www.ee.princeton.edu/~wolf/

System-Level Design with Rosetta

Perry Alexander

Department of Electrical Engineering and Computer Science
The University of Kansas

AMSTERDAM • BOSTON • HEIDELBERG • LONDON
NEW YORK • OXFORD • PARIS • SAN DIEGO
SAN FRANCISCO • SINGAPORE • SYDNEY • TOKYO
Morgan Kaufmann Publishers is an imprint of Elsevier

ELSEVIER

MORGAN KAUFMANN PUBLISHERS

Publisher: Denise E.M. Penrose
Publishing Services Manager: George Morrison
Senior Production Editor: Paul Gottehrer
Editorial Assistant: Kimberlee Honjo
Cover Design: Alisa Andreola
Composition: diacriTech
Interior printer: Sheridan Books
Cover printer: Phoenix Color Corp.

Morgan Kaufmann Publishers is an imprint of Elsevier.
500 Sansome Street, Suite 400, San Francisco, CA 94111

This book is printed on acid-free paper.

Library of Congress Cataloging-in-Publication Data
Application submitted

ISBN 13: 978-1-55860-771-2
ISBN 10: 1-55860-771-4

For information on all Morgan Kaufmann publications,
visit our Web site at *www.mkp.com* or *www.books.elsevier.com*

Transferred to Digital Printing 2009.

For my Dad

About the Author

Dr. Perry Alexander is a professor in the Electrical Engineering and Computer Science Department and Principal Investigator with the Information and Telecommunications Technology Center at The University of Kansas. He was among the original designers of the Rosetta specification language and continues to lead language design activities. He is a Senior Member of IEEE, where he has served as Vice-Chair and Chair of the Engineering of Computer-Based Systems Technical Commitee. He has published over 90 refereed publications, has presented numerous invited talks, and has won 15 teaching awards, including being named as a 2003 University of Kansas Kemper Teaching Fellow.

Contents

IV DOMAINS AND INTERACTIONS

Acknowledgments

First, I must thank my partners throughout the Rosetta development process, David Barton and Peter Ashenden. Their virtual fingerprints are everywhere in Rosetta. I have always admired Dave's intellectual curiosity and benefited from his desired to explore new things. If nothing else, Dave introduced me to Haskell and opened up new worlds for exploration. As everyone in my discipline knows, Peter is a scholar and author in a class by himself. Peter enabled us to pull the crazy things we wanted Rosetta to be into something that now actually resembles a language. I am most proud to call both Peter and Dave my friends — everyone should be so lucky.

I must also thank my friend and mentor Philip A. Wilsey, who introduced me to hardware description languages, simulation, and the sheer joy of being a scholar; Praveen Chawla and all the folks at EDAptive Computing, who have stood by Rosetta through thick and thin; Denise Penrose and Kim Honjo at Morgan Kaufmann, who have kept me pointed in the right direction and have been exceptionally patient as Rosetta has come together over the past few years; Tim Johnson, Matt McClorey, and Keith Braman, who have kept me on the straight and narrow; all of my colleagues and ITTC and KU; members of the Rosetta Committee, who have spent their own personal time reviewing and commenting on Rosetta proposals; and finally, the reviewers who have contributed to this work, Roberto Passeron, University of Trento; Grant Martin, Tensilica; Darrell Barker, Air Force Research Laboratory; Shiu-Kai Chin, Syracuse University; Greg Peterson, University of Tennessee; and Gabe Moretti, Electronic Design News.

Oddly enough, I have never met many of the people I most need to thank. I am of course referring to the research community whose work forms the basis for all that we have done. In the bibliography, I have tried to cite important works that influenced Rosetta, although I am sure there are works that I overlooked or forgot.

I thank all of the authors cited in the bibliography and apologize to anyone I may have missed.

I can't imagine where I would be without the students I've worked with over the years at The University of Cincinnati and The University of Kansas. Special thanks to all of you — Garrin, Jeni, Nick, Phil, Mark, Justin, Wang, Brandon, Kalpesh, Murthy, Cindy, Roshan, Akki, Srini, Krishna, Makarand, Amit, Arun, Iqbal, Kshama, Murali, Sat, John, Phil, Liming, Tareq, and Raj.

I've dedicated this book to my Dad and I'm quite certain that no one understands why more than my Mom. Together they have spent a lifetime enabling the people around them to achieve their goals — my sister and me in particular. This is my definition of greatness and I will never know two finer people.

Saving the best for last, I must thank my wife Pam and my son Tate, who always pretend not to notice when I wander downstairs to my office. I have two wishes for my son — that he loves what he does as much as I love what I do, and that he loves who he chooses to spend his life with as much as I love Pam.

Foreword

Now this is not the end. It is not even the beginning of the end. But it is, perhaps, the end of the beginning.

Winston Churchill, 10 November, 1942

The history of the Rosetta system-level design language spans the last decade. Indeed, the history of Rosetta mirrors the history of system-level design over that decade as well. The seminal Dallas meeting in 1997, as discussed by Perry Alexander in his Preface, crystallised the requirements for those system just emerging at that time, which also marked the beginning of the system-on-chip (SoC) revolution. The first generation of algorithmic system design tools were in active use, and researchers were exploring the meaning of such concepts as 'multiple models of computation.' The first commercial behavioural synthesis tools were struggling to demonstrate a real value to designers. The first intellectual property (IP) based SoC construction, simulation and analysis tools were trying to find an audience. And tentative experiments with C/C++ and Java as system modelling languages were a response to the perceived inadequacies of hardware design languages (HDLs).

The Dallas meeting and subsequent system-level design language (SLDL) meetings in San Jose and Italy established a set of requirements that became fixed beacons during the tumultuous system-level decade that followed. That decade saw the first generation of commercial behavioural synthesis and SoC construction tools fail in the marketplace. It also saw the system-level design language 'wars' end with a relative truce, and an industry consensus emerge on using SystemC as the imperative system design language for modelling complex hardware and software components, and for building system-level simulation models. Throughout this tumult, the need for additional capabilities in system-level design remained a constant.

In particular, the need for a notation to capture declarative aspects of system design — especially design constraints; to allow the composition of the specifications for heterogeneous subsystems; and to facilitate more formal analysis of both specification and implementation — were the subjects of many research projects — including the Rosetta project.

But Rosetta did not start out as a research project. It started as an industry-sponsored language development effort, later coming under the umbrella and sponsorship of the Accellera language-based design standards group. Interest within Accellera in Rosetta rose and fell as interest in the first phase of system-level design waxed and waned. It was fortuitous indeed that, complementing the inconsistent commitment of commercial EDA, systems and semiconductor companies to the Rosetta concept, the academic community, especially in the persons of Perry Alexander and Peter Ashenden, rose to provide a continuous level of engagement, detailed technical work, and semantic integrity in pursuing the definition of Rosetta further.

As Accellera continued to focus on the shorter term and arguably more pragmatic agenda of the commercial EDA industry and large design companies, in the form of languages such as SystemVerilog, now IEEE 1800-2005, and Property Specification Language (PSL), now IEEE 1850, it became increasingly clear that Accellera was no longer the most appropriate organisation to continue sponsoring Rosetta. Thus the language was returned to the academic community for continued specification and evolution, under the leadership of Perry Alexander. This has allowed the language to be fully specified, and has allowed the writing of this book, *System-Level Design with Rosetta*. As every new language seems to require at least one book to be written to promote understanding and further interest and use, so this milestone has now been achieved with *System-Level Design with Rosetta*.

How can we best summarise the status of Rosetta in 2006? As this book makes clear, it is a well-defined research language. There are some early tools available, and the prospect of more coming. It has a small pool of academic users, but currently very little industrial or designer usage. The availability of this book will play a big part in increasing the pool of users, in heightening interest and in encouraging more tools to emerge. What Rosetta needs above all is design use and feedback based on that use, so that it can enter a virtuous cycle — a period of rapid development and improvement based on real usage. As with any new design language, use validates the concept of it.

To use Rosetta, readers need to understand it. Here Perry Alexander's book provides a comprehensive and well-written look at the language. After a general introduction, in Section I, to system-level specification and the Rosetta concept, the book is split up into four further sections that build up the Rosetta language starting with the basic expression language, through the concepts of facets, domains, and finally, several case studies illustrating Rosetta's application.

In particular, Chapter 1 lays the groundwork for the Rosetta concept of a specification notation, illustrating this with a simple example of the basic modelling concepts, facets, and domains. It also discusses vertical and decomposition concepts in specifications and models, and ways of usage. The final section in Chapter 1 discusses how best to use the book to learn Rosetta. It will be worthwhile for the reader to read Chapter 1 carefully and to return to it for refreshing the basic concepts whenever the details in the rest of the book begin to seem overwhelming.

Section II has several chapters on the basic language constructs: value, types, expressions, and functions. Although not a formal syntax definition, careful explanation and copious examples make the base language clear. The concept of a 'facet' is the subject of Section III, and the section opens with a basic explanation of the notion and use of what will strike many readers as a new idea through simple examples. Here the concept of declarative modelling as opposed to imperative languages becomes much clearer.

In Section IV, the important notion of heterogeneous modelling domains and the composition of domains into heterogeneous specifications is explained in detail. Many examples illustrate what will be new ways of looking at modelling for most readers.

Perhaps the most important section is Section V, containing three case studies. The first line of Chapter 16 really encapsulates the idea of the whole book: "The best way to understand system-level modelling is to model systems". The first study is for an RTL specification, drawing on any knowledge of traditional RTL design that readers may have. The second study is of a TDMA wireless receiver, specifying it first with models for function, power consumption and constraints, and then illustrating refinement from specification to implementation in CMOS, FPGA, and software. The last case study is for a system-level network and access control. These three studies show the range and power of applying the Rosetta concept.

With such an introduction to Rosetta, the most important next step for the community of readers is to then start using it. This can be done even in the absence of tools or a way to reuse the Rosetta specifications. Mapping the concept of a new design and its natural language specification into the precision of a specification language such as Rosetta will be useful even without carrying it further — to clarify the specification, to clearly identify constraints, to bring out multiple heterogeneous design domains, and to demonstrate which requirements are most important. Based on such experiments, feedback on the usefulness of the Rosetta language, its constructs, and suggestions for its improvement will all be valuable for further evolution. At some point the arrival of tools will assist early users in the reuse of such specifications in formal analysis, in verification, and even possibly in implementation flows.

The arrival of the first book on Rosetta is a major milestone in the long history of the evolution of this language. I congratulate Perry Alexander on his work as demonstrated herein, and recommend this book to anyone curious about the language and wishing to study it further and apply it.

At the end of the first decade of system-level design exploration for complex SoC and embedded systems, it seems clear, as in the opening quote, that we have not arrived at the end of this topic nor even at the beginning of the end. This book truly is a marker that we are at the "end of the beginning". As Churchill said (albeit in a very different context), "Let us go forward together" into the next phase of system-level design.

Grant Martin
Pleasanton, California
April, 2006

Preface

The more I think about language, the more it amazes me that people ever understand each other — KURT GÖDEL

The last thing this world needs is one more design language. This is an odd statement to open a book about one more design language, but it is true nonetheless. What the world does need are solutions to problems that help produce cheaper, more reliable engineering solutions. Although the introduction of VHDL and Verilog revolutionized digital hardware design, today they represent venerable, established solutions to RTL design problems. The time has come to move ahead to new solutions to the new problems introduced by heterogeneity, resource constraints, rapidly evolving implementation fabrics, safety and security requirements, and system-level complexity. More of yesterday's solutions will not solve today's system-level design problems.

Rosetta was inspired by a problem whose solution demanded a paradigm shift — the design of "systems on chip," or system-level design of electronic systems. In 1997, the Semiconductor Industry Council met in Dallas, Texas to discuss potential solutions to the system-level design problem. Specifically, the industry was beginning to see the edges of what its design methods and tools were capable of supporting. The ability to fabricate systems was growing exponentially, outpacing even Moore's Law. However, the ability to design systems was growing only linearly; the causes were deemed to be lack of formal semantics in modeling languages and processes, inability to account for performance requirements during design, inability to predict emergent behaviors in heterogeneous systems; and too much reliance on simulation, among others.

During two days in Dallas in 1997, the Industry Council defined a collection of requirements for a language to support system-level design. This consensus of original motivating requirements still hangs on my wall — representation of performance requirements; precise, formal semantics; support for specification heterogeneity; and reduced reliance on simulation. Although the name has morphed from system-level design (SLD) to electronic system-level (ESL), the problem remains the same.

Almost ten years and one technology boom-and-bust cycle later, we present Rosetta. Through it all, we have tried to hold Rosetta to initial requirements while responding to the unspoken, evolving needs of system-level designers. Rosetta is formal as embodied in co-algebraic semantics supporting both simulation and modern formal analysis techniques. Rosetta represents performance constraints and supports complexity management using abstract, declarative specification. However, above all other things, Rosetta supports heterogeneous specification. This is the linchpin for effective support for system-level design decision making.

Rosetta's most important contribution to system design is heterogeneity. Heterogeneity means simply that not all elements of a specification are represented using the same semantic basis. It is embodied by the use of continuous-time semantics for analog circuits, state-based semantics for digital systems, and temporal logic for constraints all within the same design. What Rosetta provides is a language and semantics for writing such specifications and integrating them to predict system behavior. Support for heterogeneity is the enabling capability for system-level design. Without it, we are wasting our time.

Heterogeneity was woven into Rosetta's fabric from the first discussions of its semantics. Rather than establish a specification model for all systems, we have tried to establish a language framework for defining and integrating specification models. The expression language and type system provide a computation-neutral mechanism for declaring things and defining their properties. Being lazy and largely non-strict, the expression language is a means for describing calculations, rather than computations. The facet system provides means for choosing a modeling semantics, defining models using the expression language to specify properties, and composing models to define complete systems. Finally, the domain and interaction systems provide mechanisms for defining modeling semantics and interactions between models in different semantics. Because models are homogeneous and nonorthogonal under composition, we can write models for different system aspects and understand their mutual dependencies and interactions.

What this book attempts to do is provide a gentle introduction to Rosetta and the modeling techniques it supports. After working through the text, I hope that you will be able to begin writing Rosetta specifications in the base Rosetta design domains. What this book does not do is provide deep insight into Rosetta's semantics or advanced usage. It is not an attempt to replace the Rosetta standard, but instead to provide a path to understanding why the language is designed as it is. It is best to walk before we run, both as a reader and an author!

In the grand scope of electronic system-level design, Rosetta is still a young language. The standards process has started and early adopters are moving forward. There is much left to discover and more work to be done. I hope that ultimately this book will pique your interest in getting involved with the language specification and tool development efforts.

System-Level Design with Rosetta

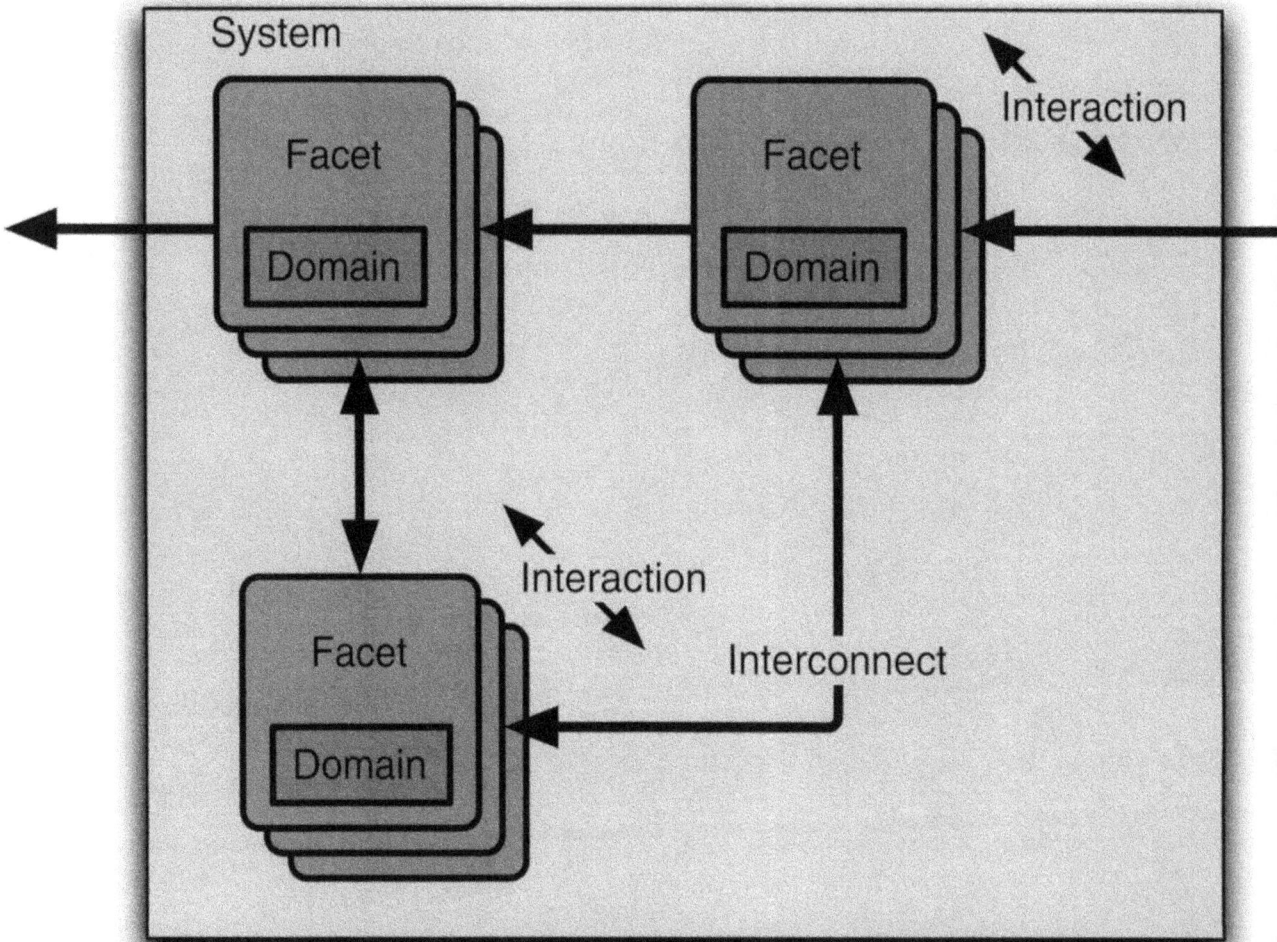

Introduction

System-level design is characterized by the need to understand the system-level impacts of local design decisions. To achieve this, a system-level designer must be able to write and compose specifications from multiple, heterogeneous domains. They must also be able to take these specifications and predict when and how they interact, effectively decreasing the intellectual distance between specification. Rosetta is a language designed first and foremost to support these activities.

Part I introduces the basic elements of a Rosetta specification by walking through a simple system-level specification. *Facets* and *domains* are introduced first as the basic building blocks of a Rosetta specification. Techniques for composing models *horizontally* to define composite systems and *vertically* to define different aspects of the same system are presented as primary techniques for heterogeneous model composition. *Interactions* are introduced in model composition as a mechanism for understanding how information from one domain impacts another. Finally, techniques for representing usage requirements as *assumptions* and correctness conditions as *implications* are presented.

After completing Part I you will have an overview of a Rosetta specification and an introduction to basic Rosetta specification concepts. Most importantly, you will have a road map for learning Rosetta and a basic understanding of how to read a Rosetta specification.

Introduction

1.1   What is System-Level Specification?

What is a system-level specification? In actuality, it is not a specification at all — no complex system will ever be defined by a single, closed-form model. It is unrealistic to believe that the designers of the Empire State Building or Saturn V had a single, closed-form solution that showed their designs would result in successful implementations. Given the relative complexity of today's systems, it is equally improbable that we will ever have single models for our next-generation design artifacts.

The answer lies in what engineers, particularly systems engineers, actually do. Cultural folklore would have us believe that engineers write and solve mathematical equations to generate designs. All engineers learn the same basic mathematics and science. A Fourier transform or set of differential equations is solved the same way regardless of discipline. If engineers simply write and solve equations, we really don't need more than one engineering discipline. However, experience teaches us that this just isn't so. To engineer good solutions, we need different disciplines with different perspectives on problems.

What a good engineer understands is when and how to apply models. After the math and science, what an engineer learns are different modeling abstractions for predicting system behavior. It isn't enough to understand calculation of a Fourier transform or solution of differential equations in isolation. What is important is knowing when and how to use these techniques — specifically, knowing when a model is applicable in addition to knowing how to solve its constituent equations.

Domain-specific models and design abstractions enable engineers to predict system behavior prior to implementation. Unfortunately, their domain-specific nature creates a veritable tower of Babel in system-level design. The design of something as common as an automobile involves elements from countless design domains — mechanical, electrical, ergonomic, economic — all with their own abstractions and all interacting. Yet, none shares a common vocabulary, making diagnosing and solving problems that involve multiple domains difficult.

The systems engineers are the general contractors for complex designs. They understand where all the parts go and how to talk to the domain engineers. Just as the general contractor must make certain a water line does not get routed through the circuit box, the systems engineer must make sure a software change doesn't result in excessive changes in power consumption. Just as the general contractor must extract information from the plumbing plans and electrical diagrams, the systems engineer must understand the software design, power consumption profile, and constraints. The systems engineer's task is characterized by the need to understand the relationship between design decisions local to a domain and system-level goals and requirements.

1.2 Rosetta's Design Goals

Many systems are simply too complex for even the best systems engineer to manage without specialized support. It is impossible to model systems with thousands of interrelated design facets manually or using general-purpose computer tools. The systems engineer needs a way to predict how a change in one component of the system affects the overall properties of the system. Having a way to express relationships and interactions between components and disciplines is a major step forward in doing this analysis.

Natural language and graphical descriptions are not reliably interpretable by computers or human engineers. Such representations are inherently ambiguous and subject to incorrect interpretation by both computers and humans. This assumes, of course, that the chosen representations can be interpreted by computers at all. Having a way to express requirements in an unambiguous, formally defined manner is central to all engineering disciplines and should be central to systems engineering.

The systems engineer also needs an unambiguous way to tell the domain specialists on the product team what needs to be designed. This is analogous to the way an architect uses blueprints to control the systems designed by plumbing, electrical, and heating, ventilation, and air conditioning specialists. Products largely fail or have to be redesigned due to an engineer designing the wrong thing because of misunderstood requirements, as opposed to designing the thing wrong because of a design error.

We have seen these issues before in other engineering and business disciplines, where they are addressed by specialized representation languages and analysis tools — precisely the motivation for Rosetta. We need a language that supports computer and human interpretation, unambiguous definition, and predictive analysis similar to those available in other disciplines. Furthermore, we need that language to address the specific needs of systems engineering. For this reason, Rosetta was developed.

How do we define a language that supports systems engineering decision making? What basic features must this language have? In answer to these questions, we established design goals leading to heterogeneous design, decomposition, abstract modeling, and formal semantics. Specifically, Rosetta was designed to support the following system-level design needs:

Multiple Domains and Multiple Semantics   Modeling in multiple domains using multiple semantics is the essence of what systems engineering and system-level design are about. We must bring together information from across engineering disciplines to understand the system-level effects of local design decisions.

Rosetta defines specification vocabularies using *domains*. Each domain defines units of semantics, a model of computation, and/or engineering abstractions for a given engineering domain. Each specification *facet* extends a domain, defining a new system specification in that domain.

Abstract Modeling   Modeling abstract, incomplete, and sometimes inconsistent requirements is key to making decisions at the system level. We cannot rely on complete, simulatable models unless we want to defer analysis until design decisions are complete. We must be able to detect and rectify inconsistencies when they occur.

Rosetta uses a declarative specification model that allows representation and analysis of abstract, incomplete information. Any specification, regardless of completeness, can be analyzed in some fashion. Furthermore, some specifications, such as constraints and performance requirements, are more easily expressed using declarative techniques. Rosetta sacrifices executability to provide these capabilities. At the system level, requirements and constraint representation is higher priority than is executability.

Vertical Decomposition and Domain Interaction   With the ability to model different system facets using different semantic models, we can now model how those system facets interact. This facilitates understanding how local changes in one domain impact the overall system. Modeling domain interactions is key to supporting true system-level design.

Rosetta defines relationships between specification domains using *interactions*. An interaction specifies a pair-wise relationship between domains by defining *Functors* that move specifications between domains. Functors are explicitly written by designers and are implicitly generated when new domains are written.

Horizontal Decomposition and Components   Structural modeling assembles components into systems and is central to virtually all engineering domains. Vertical or structural decomposition encourages model reuse, coherence, and decoupling. To support system-level design, we must model *ad hoc* interactions as well as interactions that occur through formal interfaces.

Rosetta defines structural models by instantiating facet interfaces using shared variables. Although simplistic, adding constraints to interconnections supports a rich collection of communication models. *Translators* written as a part of interactions define how information flows between facets written in different domains. Thus, components that may not share a common model of computation can be interconnected within a system.

Sound Semantics   Semantics formally define the precise meaning of specifications. We must know what our specifications mean to support analysis and synthesis, as well as sharing specifications. Semantics must support analysis techniques ranging from type checking and static analysis through simulation and emulation.

Rosetta uses a co-algebraic model for facets and a set theoretic, dependent type system. Designed with a formal semantics from the ground up, Rosetta specifications support both static and dynamic analyses.

1.3   Anatomy of a Specification

Rosetta's support for system design is best understood by examining the anatomy of a specification. We will start by modeling several different aspects of a simple register using Rosetta's basic modeling primitives, facets and domains. We will then assemble the different specifications to model a system using products and functors, critical elements of Rosetta's interaction modeling system. We will assemble components into systems to demonstrate structural modeling and system-level requirements modeling. Finally, we will discuss how Rosetta supports recording usage assumptions and implications in models.

1.3.1   Facets and Domains

Corresponding to the need for modeling different system aspects are Rosetta's *facet* and *domain* constructs. The word facet is formally defined as "one side of something many sided." A Rosetta facet is exactly that — one model of a system with many models. Each facet has an associated domain that defines vocabulary and modeling conventions for a class of specifications, thus supporting heterogeneous specification.

To illustrate the role of Rosetta's facets and domains in system-level design, we will consider a simple register. Not much of a system to be sure, but enough to allow us to illustrate the Rosetta methodology. We will start with a simple facet that describes the component's function as shown in Figure 1.1, along with a graphical representation of the device interface.

Subsequent chapters will define in detail each element of the facet model, but most elements should be familiar to anyone with software or hardware description language experience. Input and output parameters provide an interface, internal declarations provide state, and the specification body describes properties of the model. Terms in the specification body describe properties declaratively by describing constraints on parameters and internal declarations.

Figure 1.2 shows an alternative model of the same register, but viewed from a power consumption perspective. In this model, data plays no role. The model

```
facet regFcn
   (x::input word(4); z::output word(4);
    rst,le,clk::input bit)::discrete_time is
   s :: word(4);
begin
  sup: s' = if reset=1 then b"0000" else
               if clk=1 and event(clk) and le=1
                 then x
                 else s
               end if;
            end if;
  zup: z = s;
end facet regFcn;
```

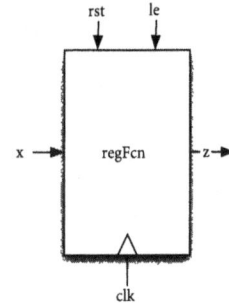

Figure 1.1 Facet description of the function of a register.

```
facet regPwr
   (rst,le,clk::input bit;
    switch,leakage::design real)::state_based is
   export power;
   power :: real;
begin
  pup: power'=power+if clk=1 and event(clk)
                      then if le=1 then switch
                                   else leakage
                           end if;
                      else leakage
                    end if;
end facet regPwr;
```

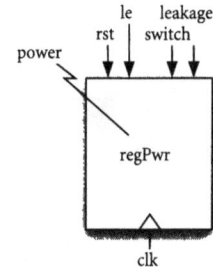

Figure 1.2 Power model.

observes the clock (clk) and load enable (le) inputs to determine if the device switches and updates the consumed power value appropriately. Two design parameters, leakage and switch, allow the specifier to update power calculation constants associated with power leakage and power used when the component switches. The consumed power is observed by examining the exported value regPwr.power at any time during system operation.

Finally, Figure 1.3 provides a simple cost model for using the register in a system. This is a trivial model providing a constant cost value for each register. In a fashion similar to that of the power facet, the exported regCost.cost value can be used to observe register's cost. More complex cost models taking into account device width or implementation technology could easily be developed.

The three register models are quite similar in structure and appearance. Closer examination reveals two fundamental differences among them—interfaces differ from model to model, and each facet is defined with its own domain. Interfaces vary based on the interface of the model represented. The functional model interface looks like one would expect with data inputs and outputs, control

```
facet regCost :: static is
  export cost;
  cost :: real is 0.02;
begin
end facet regCost;
```

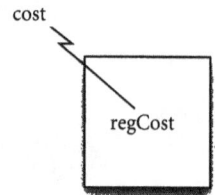

Figure 1.3 Cost model.

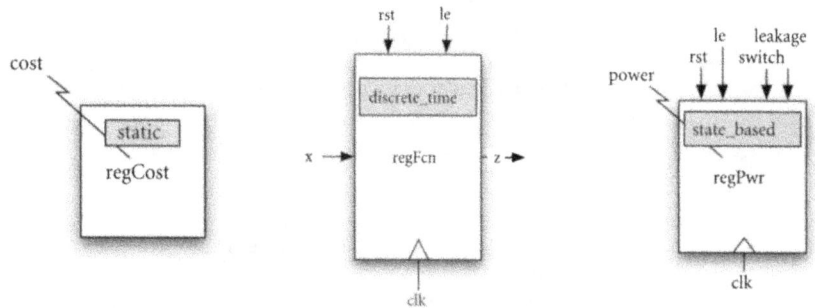

Figure 1.4 Register facets with domains exposed.

inputs, and a clock. The power model is written independently of data inputs and outputs, but adds parameters for adjusting power calculation constants. The cost model is completely independent of inputs and outputs and thus has no parameters. In both the power and the cost models, observable properties are not a part of the interface, but are directly accessed as exported declarations. This distinction is subtle, but important. Cost and power are not system inputs and outputs like data and control, but rather are properties observed over the system as a whole.

Each register model is defined using a different *domain*. Figure 1.4 shows each facet with its associated domain highlighted. A facet's domain defines a vocabulary and model-of-computation for the model, allowing each Rosetta facet to use a different semantics. This is critical to system-level design because it allows us to use a semantics appropriate to each model, rather than forcing models to a single semantics. The systems engineer does not force all models into the same semantics, but instead integrates models from different semantics. The key to Rosetta's system-level design support is this ability to use appropriate semantics for each model and integrate the result into a system-level specification.

1.3.2 **Vertical Decomposition**

The actual register is not any one of the individually defined models, but rather is a *system* described by those models concurrently. Additional models could easily be written in a similar fashion modeling such characteristics as electromagnetic interference (EMI), analog behavior, setup and hold, and packaging properties. To model the entire register, we must assemble the register facets into a composite register model. To do this, we will use Rosetta's *product* operator to assert that the models all describe the same register.

The product operator is rather innocent looking, but is exceptionally powerful in application. Quite simply, if F_1 and F_2 are both facets, then $F_1 * F_2$ is a new facet that embodies both of the original facets. Thus, the Rosetta declaration:

```
reg(x::input word(4); z::output word(4);
    rst,le,clk::input bit;
    switch,leakage::design real) :: static is
  regFcn(x,z,rst,le,clk) * regPwr(rst,le,clk,switch,leakage) * regCost;
```

puts all the register models together into a single model, reg, that embodies each of the original models. Whenever reg is used in a specification, all facets comprising that model are present and must be consistent. Figure 1.5 graphically shows the model resulting from the product as a collection of facets "stacked," representing *vertical decomposition* of the single component.

The model in Figure 1.5 is somewhat different than the pure product model for reg. The preceding reg product reduces each model to a *least common semantics*, that being a model in the static domain. This defeats the purpose of using multiple domains. What we want is to move models to the most appropriate domain for the kind of analysis needed. The following register model does this using *functors* on facet models:

```
reg(x::input word(4); z::output word(4);
    rst,le,clk::input bit;
    switch,leakage::design real) :: discrete_time is
  regFcn(x,z,rst,le,clk)
  * state_based_discrete_time.gamma(regPwr(rst,le,clk,switch,leakage)
  * static_discrete_time.gamma(regCost);
```

```
reg(x::input word(4); z::output word(4);
    rst,le,clk::input bit;
    switch,leakage::design real) :: discrete_time i
  regFcn(x,z,rst,le,clk)
  * state_based_discrete_time.gamma(regPwr(rst,le,c
  * static_discrete_time.gamma(regCost);
```

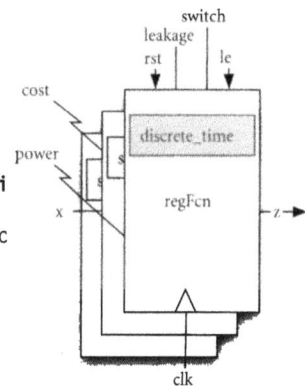

Figure 1.5 Register models composed using a product and functors.

This new reg model creates a model in the discrete_time domain and the two gamma functors move the regPwr and regCost models to the discrete_time domain. The product is then formed in that domain without loss of design abstractions used in the functional model. This is key to Rosetta design and usage. Information is moved from one modeling domain to another to preserve or introduce design abstractions useful for analysis. Together, functors and products define the central features of the Rosetta *interaction system*.

Another important aspect of the Rosetta interaction system is the *facet combinator*. The product operation assembles individual facets into composite system models. The facets still exist within the product as separate, orthogonal models. In effect, the product packages up individual models into a single structure and can still be referenced as individual models. Facet combinators take facet constructions like products and compose their properties into a single facet model, allowing the systems engineer to express interactions.

If we want to write more sophisticated power consumption models that take into account actual data processing and processor behavior, we can write a combinator between the functional and power models. Such a combinator would be used in a specification as follows:

```
regFcnPwr(x::input word(4); z::output word(4);
   rst,le,clk::input bit;
   switch,leakage::design real) :: discrete_time is
   fcnPwrComb(regFcn(x,z,rst,le,clk),
      state_based_discrete_time.gamma(regPwr(rst,le,clk,switch,leakage)));
```

This definition uses the gamma functor to move the power model into the discrete_time domain and then the fcnPwrComb to compose the resulting model with the register's functional model. In this new model, the power will be an observed property in the same manner as the original model. Thus, regFcnPwr.power will be a defined, observable value that accounts for the register's functional behavior. The result is a much more accurate model.

1.3.3 Horizontal Decomposition

Among the most commonly used engineering design techniques is *structural decomposition*, where a system is represented or implemented as a collection of subsystems. The impacts of decomposition in this manner are obvious when considering the reduction in complexity of models and components. In addition to vertical decomposition, Rosetta's facet modeling system supports structural decomposition, where a system is defined as a collection of interconnected components.

Figure 1.6 shows a simple structural model using the reg model developed previously. The form of this model should be familiar to those experienced with hardware description languages. Shared declarations are used to communicate between the various system components. All system components are modeled similarly to the register, with various models representing different system facets.

```
facet controller
    (rst,clk,le::input bit;
     leakage,switch::design real;
     o::output bit) :: state_based is
  export power;
  power::real;
  x,z::word(4);
begin
  r0: reg(rst,clk);
  c1: ns_logic(z,x);
  c2: output_logic(z,o);
  p1: power' = power +
        r0.regPwr.power +
        c1.ns_logicPwr.power +
        c2.ouput_logicPwr.power;
end facet controller;
```

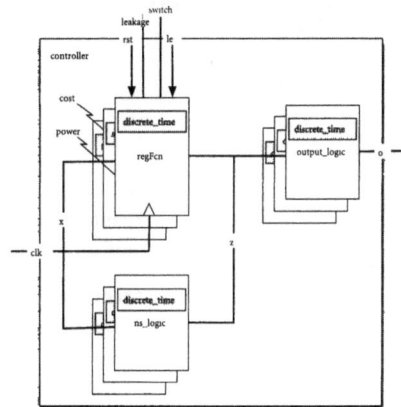

Figure 1.6 Using the register model as a system component.

A distinctive feature of this structural model is the calculation of system power consumption from component power consumption values. Here, the power models associated with each component are extracted, then power values extracted and added to the cumulative power consumption value. We are mixing structural models with property models to calculate system-level performance values. The power consumption of our system can be observed as controller.power, allowing the systems engineer to observe changes in system power consumption resulting from local design decisions.

1.3.4 Vertical Decomposition — Revisited

Of course, the hierarchy of Rosetta models is not strictly one level deep. New models may be defined and composed with the structural controller model in exactly the same manner as for the original register model. It is quite common for a system to be described using multiple architectures. One common example of this occurs when an embedded system is described as a process diagram and simultaneously in terms of its implementation architecture. Figure 1.7 shows such a situation for our controller example.

Figure 1.7 graphically shows two architectures for our controller system. One is the controller architecture developed previously. The other is an implementation architecture involving a central processing unit (CPU) and a memory block. The first Rosetta definition forms the product of these models to indicate that they both describe the system we are constructing:

```
facet regImpProd :: discrete_time is reg * cpuArch;
```

The regImpProd model describes a system that functions like reg and whose implementation is structured like cpuArch. However, the actual implementation

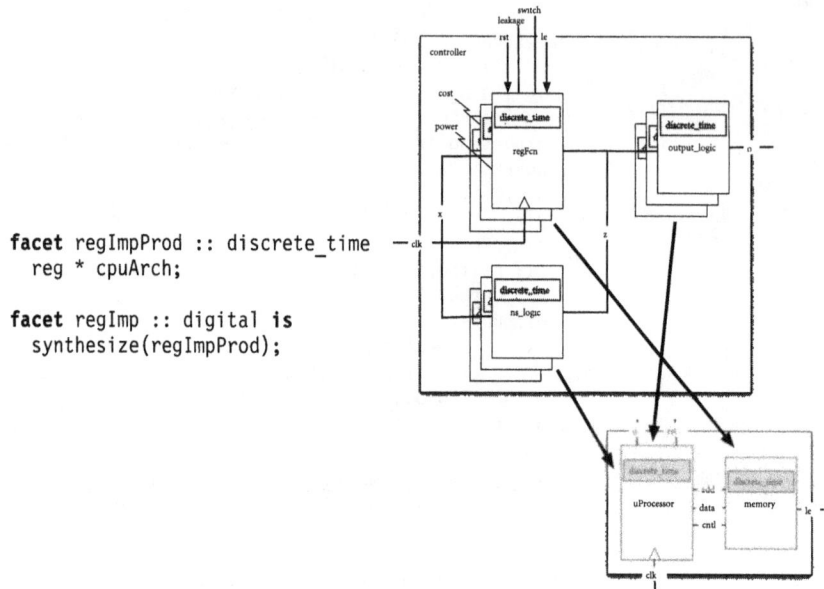

```
facet regImpProd :: discrete_time
  reg * cpuArch;

facet regImp :: digital is
  synthesize(regImpProd);
```

Figure 1.7 Implementing the register model.

is not yet known — only the fact that our final implementation exhibits both sets of properties is.

The second Rosetta definition is a facet combinator that synthesizes an implementation from the two models:

```
facet regImp :: digital is synthesize(regImpProd);
```

The regImp fact is formed by synthesizing a model from the functional and structural models with the facet combinator synthesize. One should quickly recognize that the synthesize combinator is not at all a trivial function to write, and Rosetta provides no silver bullet for addressing this problem. However, it does allow us to define relationships between models that support verification and traceability in the systems design process.

1.3.5 Usage Requirements and Implications

Once a system has been designed and fielded, it is easy to lose track of design assumptions that govern its use and testing. Such information is virtually impossible to glean from an existing system, yet it is vital to the systems engineer. When a system is initially fielded, the systems engineer must understand preconditions for and implications of proper operation. After a system has been fielded and must be updated or replaced, the same information is vital for understanding design context. Knowing what operating conditions were anticipated for a system (as well as what should be observed during operation) dramatically simplifies the system design task.

Rosetta provides a *component* construct for recording operating preconditions and implications. The component construct is in all ways a facet and can be used anywhere a facet is used. However, the component provides mechanisms for specifying preconditions and implications for using the represented system.

Figure 1.8 shows a component defined around the functional register model. The definition of the register remains unchanged. In fact, the component model references the `regFcn` model as its definition. The **assumptions** section places usage requirements on the component. In this case, the only specified usage requirement is that the frequency must be less than 5 GHz. This condition must be satisfied by any system using the component before proper operation can be assured. If we included this component in the previous structural model, we would be obligated to assure this condition based on known facts about the operating environment. The **implications** section defines correctness conditions on the component. The systems engineer must be able to infer these conditions from the system requirements and usage assumptions.

The component structure provides a semantics for expressing and verifying usage conditions and implications, but does not place strict requirements on the verification mechanism. This is due to the wide variety of verification mechanisms used in systems design, ranging from simple observations to formal verification. Rosetta does provide a justification language for recording verification support, but it is used only for recording support and is not a verification language.

1.4 Learning Rosetta

Learning Rosetta is as much learning a methodology as learning a new modeling language. Although the register example is quite small, it demonstrates many of Rosetta's features that enable approaching design in new ways. From support

```
component
  regComp(x::input word(4);
          z::output word(4);
          rst,le,clk::input bit)::discrete_time is
begin
  assumptions
    freq(clk)<5e9;
  end assumptions
  definitions = regFcn;
  end definitions
  implications
    le=0 implies z'=z;
  end implications
end component reg;
```

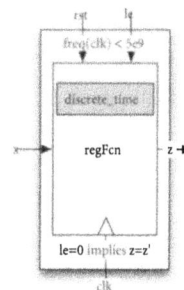

Figure 1.8 Register component defining a frequency limit and correctness condition.

for heterogeneous modeling to using facet combinators to generate models, the modeling process becomes a part of the model itself.

As we step through Rosetta modeling capabilities in detail, we will place them in the larger context of supporting system-level design. The remainder of the text builds Rosetta up from the bottom, starting with primitive specification capabilities and working toward functors and specification composition.

Part II presents the Rosetta *expression language*, the underlying language for writing all Rosetta specifications. The expression language provides a rich type system and expression definition mechanisms for declaring and defining properties over specification structures. The expression language is a lazy, nonstrict functional language with no side effects. It supports defining mathematical properties as opposed to writing imperative programs.

Part III presents the Rosetta *facet language*, the language for defining basic models. The facet language supports defining a collection of terms over parameters and local variables, and identifying a modeling domain. All Rosetta models are based on facet semantics, thus the facet language is the fundamental construct for writing Rosetta specifications. Packages and components are specializations on facets that provide standard means for structuring specifications.

Part IV presents the Rosetta *domain and interaction language* and the *facet algebra* for defining modeling domains, composing specifications, and defining various interactions between models. The domain language defines models of computation, modeling vocabularies and elaboration semantics for specifications. The interaction language packages functor, combinator, and translator specifications into constructs that define domain interactions. The facet algebra defines a collection of operations useful for combining specification models in the presence of interactions.

Finally, Part V presents three case studies of Rosetta system specification. The first case study examines a simple register-transfer level (RTL) system. It serves as an introduction to system-level modeling concepts using an example that is familiar to most designers. The second case study examines a time division multiple access (TDMA)-based telecommunications system looking at relationships between design alternatives and power consumption. It examines the TDMA-based system first from a behavioral specification and then expands to demonstrate structural modeling capabilities. The final case study examines simple security requirements in a networking environment, looking at moving a mobile system from one network infrastructure to another. The three case studies involve different issues in different systems, but are all addressed through heterogeneous modeling, model composition, and constructing interactions between models.

By stepping through each part in sequence, basic Rosetta specification techniques can be mastered. In Part II, the reader should concentrate on learning how to express mathematical properties using the Rosetta expression language. In Part III, the reader should concentrate on learning how to use facets and domains to specify system properties around domain-specific vocabulary and computation modeling elements. The expression language is used to define properties, facets are used to encapsulate properties, and domains are used to introduce computation models. In Part IV, the reader should concentrate on specifying different modeling vocabularies, computation models, and relationships between models. Domains are constructed to define basic computation models while interactions are used to model how different specifications interact. In Part V the reader should concentrate on learning how to define systems and specify analysis and synthesis tasks related to multiple design facets. The case studies attempt to provide examples of classic Rosetta specification techniques and provide a departure point for future specifications. Finally, the bibliography lists a number of works that have influenced Rosetta's design. They include work from formal methods, language semantics and type theory, hardware specification languages, and programming languages.

One thing this text does not try to do is define a complete, unambiguous syntax for the full Rosetta language. Regardless, there is a need to describe syntactic conventions throughout the book. Table 1.1 provides a list of syntax description conventions used throughout the book. Where possible, traditional notations are used.

Table 1.1 Table of syntax description directives

Syntax	Meaning
e	Expression meta-variable
$e_1 \mid e_2$	Choice between e_1 and e_2
keyword	Rosetta keyword
$[\![e]\!]$	0 or 1 e
$[\![e]\!]^*$	0 or many es
$[\![e]\!]^+$	1 or many es

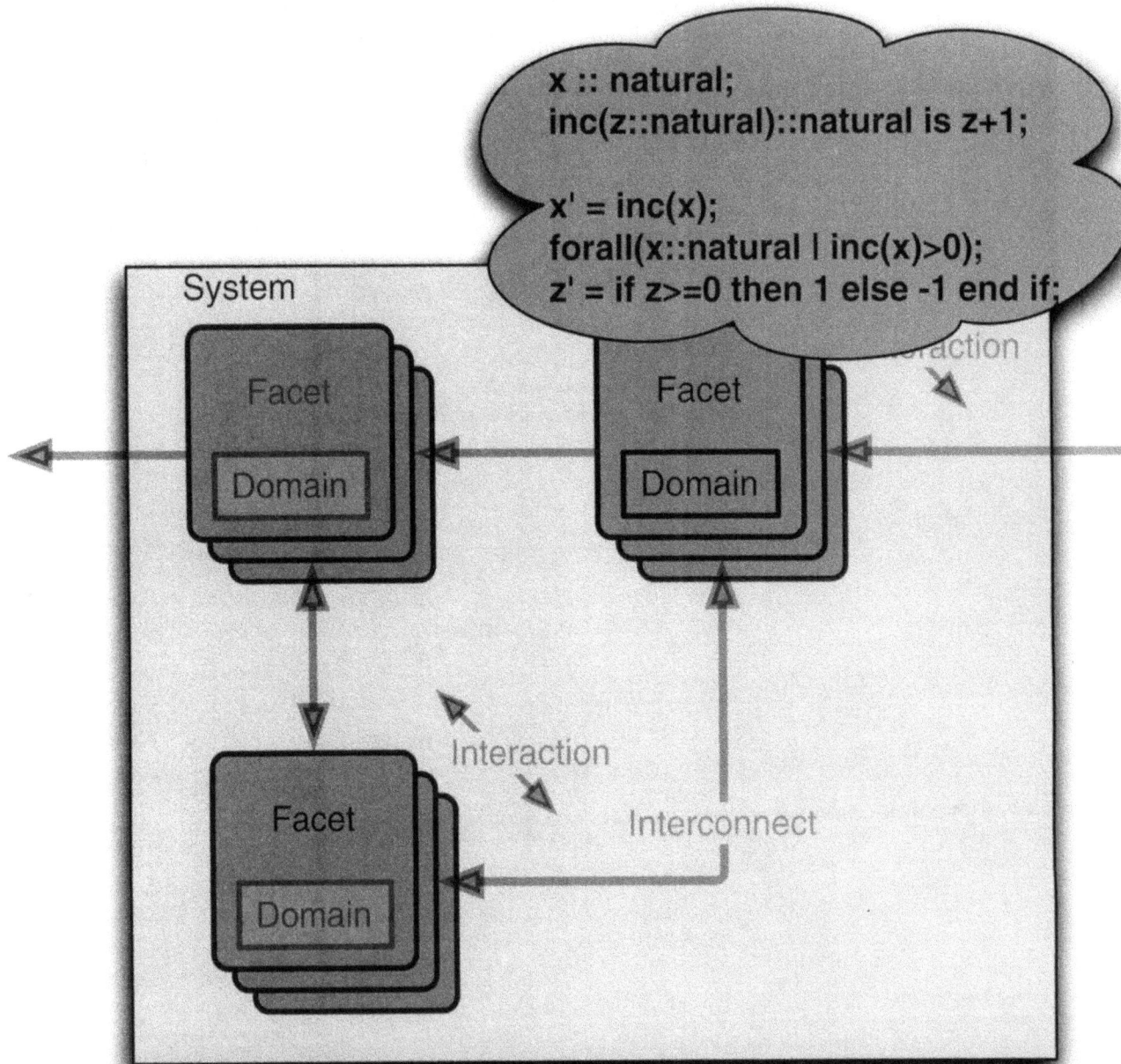

The Expression Language

Before writing specifications, we must define a language to describe system properties. In Rosetta, system properties are describe with respect to *items* that represent units of specification. All items have *types* and *values*. By describing relationships between different items through their types and values, system properties can be defined.

Part II describes the Rosetta *expression language* used to declare items comprising a specification and define properties over those items. The expression language is used to declare items, assign types to items, and define properties among items. It is declarative, allowing specifiers to define properties. In contrast to writing programs that indirectly exhibit desired properties, declarative techniques allow the specifier to state specific properties directly. Furthermore, these properties can be incomplete and need not be executable.

After completing the chapters in Part II, you will understand how to declare items, use elemental and composite types, define and use functions and higher-order functions, and define new types.

Items, Values, Types, and Declarations

The heart of any Rosetta specification is a collection of *items* that represent observable quantities associated with a system. They represent a collection of things that, if properly described, provide a precise system model that can be reasoned about to predict behaviors. By defining item properties and relationships between items, a Rosetta specification defines expectations on their collective behavior and the system they represent.

An item is defined by a declaration that associates it with a type and optional value. The *type* represents the collection of values the item can legally assume. Rosetta types are formed from sets using traditional set operations, comprehension and extension. The optional *value* makes an item a constant by associating a specific value with it. Values are defined in the traditional sense as terms that are in irreducible, normal form.

Like type systems in programming languages, the Rosetta type system specifies constraints by associating a type, like integer or **set**(character), with an item. Unlike traditional programming language type systems, Rosetta types may be defined by comprehension using a property to filter an existing type. Thus, a Rosetta type asserts properties on an item by restricting it to a set of values or defining a property the item must have to be of a particular type. Using types to specify properties in this way is a critical part of Rosetta specification. The first step in this process is understanding item declaration and how types and values are associated with an item.

2.1 Labels, Values, and Types

All Rosetta definition structures are represented semantically as items with a label, type, and value properties. The item label provides a name for the item that is used to reference it in the specification. In the Rosetta expression x=y+1, the labels x and y refer to the declared items with those respective labels. The item value specifies a value associated with the item. In the Rosetta expression x=y+pi, the value of the item labeled x is constrained to be equal to the value of the item labeled y plus the value of the item labeled pi. The item type is a set that specifies a collection of values the item can legally take. Given an item v whose type is the set T, the value associated with v must always be an element of the set T.

2.1.1 Labels

A Rosetta label represents a legal name for a Rosetta item. The rules for writing labels in Rosetta are similar to those of other languages. A label must start with a letter, which may be followed by any number of letters and digits. Single underscore characters (_) may be included to enhance readability, but they may not occur at the beginning or end of a label. Examples of legal labels are:

```
a N timeout MaxTravel Min_Travel port1 port_2a
```

Examples of illegal labels are:

```
3a              // starts with a digit
_id             // starts with underscore
packet_         // ends with underscore
sequence__0     // two successive underscores
```

The case of characters in labels is not significant. The following labels are all treated as the same:

```
pressure Pressure PRESSURE
```

While we have a great deal of flexibility in choosing item labels to aid readability of specifications, there are some labels we cannot use, since they are reserved as keywords. These keywords provide the syntactic structure of a Rosetta model and are listed in Table 2.1. Note that, as in item labels, case is insignificant. In this book, we follow the convention of setting keywords in boldface to distinguish them from item labels.

2.1.2 Values

A Rosetta value is a structure or term in normal form that cannot be further evaluated. Although not all normal forms are values, all values are normal forms. The simplest Rosetta values represent numbers, characters, and booleans. Examples of such values include the numbers 1 and 3.1415, the characters 'a' and '@,' and the boolean constants **true** and **false**. Such values are elemental values and cannot be decomposed. We form composite values from other values like the constant sequence [1,2,3,4], the constant set {high,low,unknown}, and the multiset constant {*1,2,3,1,2*}. The sequence of sets [{},{1},{2,3}] is also a composite value because it is formed from other values.

All elements of a specification are items that have associated types and thus potential values—a function item has an associated *function value*, a facet item has an associated *facet value*, and an element of a constructed type has a *constructed value*. A function value is an anonymous function that encapsulates

Table 2.1 Rosetta keywords

and	data	facet	let	or	then
array	definitions	false	library	output	top
assumptions	design	forall		of	translators
as	div	from	max		true
	dom	functors	min	package	type
be	domain		mod		
begin		if	multiset	ran	use
between	element	implications		rem	
body	else	implies	nand	ret	var
bottom	elseif	in	nmax		
	end	infinity	nmin	sel	where
case	enumeration	input	not	sequence	with
combinators	exists	interaction	nor	set	
component	export	interface	null	sharing	xnor
constant		is		sub	xor
				subtype	

an expression with a collection of parameters. Rosetta function values are pure functions in that they may only specify a functional relationship between their parameters and output values. A facet value represents the basic unit of Rosetta specification and is associated with facets, domains, components, and packages. Finally, constructed values result from the application of constructors associated with constructed types. Rosetta constructed types provide mechanisms for users to define completely new types by specifying type constructor functions and observer functions for a collection of values.

2.1.3 Types

A Rosetta *type* defines a collection of values that constrains the value of an item. Within the Rosetta language system, all types are values and appear as sets. Technically, not all types are sets, but set operations are always available on them. This fact and its implications will be discussed extensively in subsequent chapters. For now it is sufficient to understand that an item's value must be an element of its type.

If one Rosetta type is a *subtype* of another, then it follows that every element of the subtype is a member of the supertype. This is precisely the definition of subset. It follows that Rosetta subset relations also indicate subtype relationships. The following relationships are examples of this principle:

```
integer =< real  // integer is a subtype of real
integer < real   // integer is a proper subtype of real
```

```
real >= integer   // real is a supertype of integer
real > integer    // real is a proper supertype of integer
```

Rosetta item values must always be taken from their associated types. This fact is the heart of Rosetta type and constraint checking. As we will see, using a sophisticated and flexible type system allows assertion and verification of complex properties in a compact and readable form.

2.2 Item Declarations and Type Assertions

Item declarations occur in declarative regions, parameter lists, functions, and **let** forms where new variables and constants are required. Each item declaration creates a new item with a label and must be annotated with a type assertion. A Rosetta type assertion uses the assertion operator ":: " to associate a type with an item. Whenever we use the notation x::T, we are asserting that the value associated with the item labeled x is a member of the type T. All declarations must include a type assertion, but type assertions may be used to indicate types anywhere in a Rosetta specification. Thus, type assertions are used to assign types to new items and to specify types for items in expressions.

2.2.1 Item Declarations

When the type assertion x::T appears in a declarative region such as a facet's local declarations, a **let** clause or parameter list, it represents an item declaration or simply a declaration. A declaration of this form adds a new item labeled x to the current lexical context with the assertion that x is a member of type T.

Each declaration achieves three results: (i) it creates and labels a new item in the current scope; (ii) it assigns a type to the item, defining potential values for the item; and; (iii) it may bind a value to the item. If a declaration does not bind a value to an item, the item is considered a *variable*. If a declaration does bind a value to an item, the item is considered a *constant*.

An example of a Rosetta declaration that defines an item named mean is:

```
mean :: real is 13.5;
```

In this declaration, mean is the label of the new item and provides a name used to refer to the item. The type assertion operator assigns a type to the newly declared item. In this example, the type specified is real, requiring that any value associated with the item must be a real number. The type specification is followed by an optional value constraint specified by the **is** clause that binds the item's value to a constant. In this example, the value of mean is equal to the constant value 13.5.

The value specified by the **is** clause is not an initialization value. The **is** clause asserts that the declared item's value must always be equal to the specified value. Thus, the value of mean will always be 13.5 in the scope of this declaration.

Such declarations are referred to as *constant items* or simply *constants* and can be identified by the presence of the **is** clause in the declaration.

We define a *variable* by excluding the **is** clause and value from an item declaration. Without this constraint, the value of the item is allowed to vary among all possible elements of its type. For example, the declaration:

```
variance :: real;
```

defines an item named `variance` whose type is `real`, but whose value is not constrained beyond the type in the declaration. When the label `variance` appears in the scope of this declaration, it refers to the value associated with `variance`. Unlike a constant declaration, the variable item's value will be constrained elsewhere in the specification, rather than the declaration. Such declarations are referred to as *variable items* or simply *variables* and can be identified by the absence of the **is** clause in the definition.

A variable or constant may also include a constraint in the form of a **where** clause:

```
mode :: real where (mode > −5.0) and (mode < 5.0);
```

This declaration specifies that `mode` is an item of type `real` that must be between the values −5.0 and 5.0. The item's value may vary like a variable, but must always satisfy the **where** clause predicate.

In general, all Rosetta item declarations take the following form:

```
items :: T ⟦ is e ⟧ ⟦ where p ⟧ ;
```

where *items* is a comma-separated list of one or more labels, *e* is an expression, and *p* is a predicate.

In the item declaration, *T* constrains the potential values of the items to the contents of a type. If the optional **is** clause is present, the new item is a constant item whose value is defined by the expression *e*. Otherwise, the new item is a variable item. If the optional **where** clause is present, then *p* must hold for all values the item takes on. If the item is constant, the **where** clause defines additional properties that must hold for the constant value. If the item is variable, the **where** clause defines additional properties beyond those defined by the type.

For example, the function declaration:

```
triangulate(x,y :: position) :: position;
```

declares two parameters of type `position` that may be used in the function body associated with `triangulate`. The declarations create the local parameters and the appropriate type assertions. In parameter declarations, the **is** and **where** clauses are not allowed.

Similarly, the facet declaration:

```
facet parity(x::input bit; z::output bit)::state_based is
  s::bit;
begin
  s' = x xor s;
  z = s';
end facet parity;
```

declares two parameters, x and z of type bit, representing component input and
output, respectively. Like function parameters, the **is** and **where** clauses are not
allowed in facet parameters. However, a parameter kind that specifies properties
is allowed. Here, the **input** and **output** kinds provide directional information for
parameters that will be used as ports. In the declarative region, s::bit defines
a variable item and asserts that it is of type bit. These declarations are visible
throughout the specification body along with the constraint that their values must
come from the bit type.

The function and facet definition forms used thus far are simply syntactic sugar
for a traditional declaration and can be expressed using the traditional declaration
syntax. The function definition specifies a type constraint using the type assertion
operator to define the function's domain and range elements as type position.
The facet definition contains a type assertion operator asserting the facet is of type
state_based and identifying the domain that forms the basis for this specification. This type assertion can be observed following the facet's parameter list and
is no different than those defining parameters or variables. It asserts that the facet
parity is a member of the state_based facet type.

EXAMPLE 2.1

Basic declarations

```
package declaration_examples :: static is
  w :: integer;
  x :: real is 2.4;
  y,z :: string;
  a :: integer where a>5;
  b :: real is 2.6 where b > 1.0;
end package declaration_examples;
```

This example package defines six new items and groups them together into
a reusable package structure. w defines a variable item of type integer while
x defines a constant of type real whose value is specified as 2.6. The next
declaration defines two variable items, y and z, both of type string. If an **is** clause
were used in the string declaration, the result would be two constant items with
the same value. Specifically:

```
y,x :: string is "Hello_World";
```

results in two items whose values are both the constant string "Hello_World".

The final two declarations use the **where** clause to define constraints on
possible values. The declaration of a is constrained to allow only values greater
than 5 while the declaration of b must be greater than 1.0. In the latter declaration, the constraint can be verified immediately because b's value is known. In the

former, the constraint must be placed in the context of another expression to be verified. ■

2.2.2 Type Assertions

The type assertion operator may also be included in expressions when specifiers need to attribute additional type information to an item. In the definition:

```
z = x::integer + y::integer;
```

the type assertion operator is used to add type information to an expression, not to declare a new item. If type assertions are omitted:

```
z = x + y;
```

the Rosetta type system is used to determine the types of x and y. The type assertions tell processing tools to add an assertion that x and y are of type integer rather than invoking the type inference system. Such uses of the type assertion relation are simply called type assertions and differ from declarations in that they do not create new items.

Type assertions must be used with care, as nothing prevents assertions from introducing inconsistencies in specifications. However, they are quite useful when an item's type cannot be inferred automatically, when a prior type inference activity requires documenting, or when the type inference process needs a hint to proceed.

2.3 Universal Operations

Table 2.2 defines defines operators defined over all Rosetta values and types. Type assertion and subtype and supertype relations support forming type assertions and declarations, and asserting subtype and supertype relationships. Because the type of these relations is boolean, the use of **top** does not introduce problems.

Table 2.2 Operators defined over all Rosetta items

Operator	Syntax	Expression type
Type assertion	x :: T	T
Set membership	x **in** T	boolean
Subtype and proper subtype	S =< T, S < T	boolean
Supertype and proper supertype	S >= T, S > T	boolean
Equality and inequality	A = B, A /= B	boolean
Equivalence	S == T	boolean

We have seen how the type assertion operator declares items and makes type assertions. Note that the type of the type assertion operator is not boolean, but is the asserted type. Although this may appear odd, it allows embedding of type assertions in other expressions. For example:

```
1::integer - 1::real;
```

is not legal if the type assertion operation is boolean. Even if it were legal, it does not mean what we intended to say. If the type of the type assertion operation 1::integer is integer, then traditional type checking works as expected. Along the same lines, if 1::integer evaluates to 1, then evaluation also behaves as expected. Specifically:

```
1. 1::integer + -1::real
2. == 1 + -1::real
3. == 1 + -1
4. == 0
```

The set membership operator, x **in** T, provides the canonical element operation from set theory. The operator is true if its first argument is an element of its second argument. From the user's perspective, the set membership operation is equivalent to the type assertion operation because types look like sets. In general, types are not sets, but always look like sets to the specifier. Unlike type assertions, the type of the set membership operator is boolean, allowing it to be used in terms and other boolean statements.

The subtype and supertype relations define relationships among types, again treating types like sets. We will see later that subtype relations correspond with subset relations and supertype relations with superset relations.

Equality and inequality define traditional equality relations. If x=y is asserted, then the value of x is the same as the value of y. This is not the same as assignment, where x is assigned the value of y after executing the statement. In the C and C++ programming languages, the following notation would be used to increment the value in variable a, assuming that a is an integer:

```
a = a + 1;
```

This expression is inconsistent in Rosetta, as it asserts that the integer value associated with a is equal to its successor. In a traditional programming language, the semantics of the statement are that the value of a following statement execution is equal to the value of a prior to execution plus 1. Rosetta allows a similar concept when state and state change are a part of the computation model in use. Specifically:

```
a' = a + 1;
```

asserts that a in the next state is equal to a in the current state plus 1. Rosetta simply makes the distinction between current and next state visible to the specifier, rather than hiding it in the semantics of assignment. This allows Rosetta specifications to use different next-state definitions based on the computational model being used.

When we make the assertions:

```
A = 1
(a^2 + 2*a*b + b^2) = (a+b)^2
sqrt(x)*sqrt(x) = x
A/= 1
```

we are defining axioms over the values and items involved. Although this is true for any Rosetta term, mistakes tend to be made when thinking of equality and assignment as the same operation. Rosetta has no assignment operator and thus equality must always be viewed as a relation between values instead of assignment.

Rosetta also provides an equivalence operator that is semantically identical to equality. The distinction is that equivalence binds after virtually all other operations, avoiding confusing parentheses. As an example, consider the following two equivalent terms:

```
(f(x) = b) = (g(y,z) = d)
f(x) = b == g(y,z) = d
```

Dropping the parentheses from the first term results in a semantics quite different than intended. To see the distinction, we can make the left associativity of equality explicit using parentheses:

```
(((f(x) = b) = g(y,z)) = d)
```

Making equality right associative or making equality bind later introduces different problems. However, using equivalence solves the problem. Because equivalence binds last, explicitly parenthesizing the second term above results in exactly the desired semantics:

```
(f(x) = b) == (g(y,z) = d)
```

Parentheses can be eliminated because of the low precedence of equivalence. The resulting specification is more intuitive and easy to write. The equivalence relation is used extensively to provide definitions for functions and terms. Throughout this and subsequent chapters, equivalence is used extensively to provide definitions. Rosetta specifications also use equivalence to provide definitions where other mechanisms limit expressiveness.

Expressions 3

The Rosetta expression language is used to define properties involving
Rosetta items. The simplest Rosetta expression is an *atomic expression* consisting
of a literal or an item name and forms the leaves of the expression tree. *Function*
and *operator applications* define operations over other expressions by instantiating
function values with actual parameters. The *if* and *case expressions* define selection
operations using boolean conditions and set membership, respectively. The *let
expression* defines local variables over expressions. Because all expression types
other than atomic expressions are defined recursively over expressions, *compound
expressions* are formed by nesting expressions of all types.

All Rosetta expressions from atomic to compound have associated types.
Determining types is instrumental to analyzing specifications for type safety, a
critical aspect of system correctness. Due to the richness of the Rosetta type system, typing and type checking are particularly critical to successful design activities. To determine the type of an expression, Rosetta uses a combination of
declared and asserted type information with rules for each expression type. The
types of atomic expressions can be learned from their declarations. Similarly, the
types of function and operator applications can be learned from their declarations and parameter values. Special rules exist for inferring the types of *if*, *case*,
and *let* expressions from their sub-expressions.

3.1 Atomic Expressions

An *atomic expression* is the simplest expression consisting of a value or an item
label. Such expressions are called atomic because they cannot be reduced to a simpler form and have no identifiable parts. Examples of atomic expressions include:

```
1           // Value of 1 in base 10
-16         // Value of -16 in base 10
2\1111\     // Value of 15 in base 2
123e2       // Value of 12300 in base 10
2\1111\e1   // 11110 in base 2
```

```
'a'        // Character a
a          // Item named a
"Hello"    // String value
Hello      // Item named Hello
true       // Boolean true value
```

The examples shown demonstrate several mechanisms for specifying atomic items ranging from number values and string literals to item names.

An item label used as an expression refers to the item it names. Thus, the type associated with an item label is the type of its item's declaration. Because each item must be declared and each declaration includes a type, the type of a declared item is always known.

The type associated with a literal value is discussed extensively in Chapter 4. For now it is sufficient to understand that a literal's type is its most restrictive legal type. For example, the number 3 is an element of several types, including `complex`, `real`, and `posreal`. When appearing in a specification, its type is `posint` because no subtype of `posint` contains 3. However, when performing type checking, 3 can be treated as `posint` or any of its supertypes.

3.2 Function Application

A *function application* is simply the instantiation of a function with a collection of actual parameters. For example:

```
cos(x)
```

is the application of the cosine function to an item, x.

Function applications are specified using the following form:

$$f(e_0, e_1, e_2, ..., e_n)$$

where f is the name of an item whose type is a function and e_0 through e_n are expressions representing the function's actual parameters; e_k may be any expression of the same type or a subtype of its associated formal parameter in the definition of f. Thus, actual parameters may be function applications, sequence references, literals, or any other expression satisfying type conditions. Examples of function applications include:

```
inc(1)        // Increment the value 1
add(1,2)      // Add the values 1 and 1
inc(x)        // Increment the value of x
add(1,inc(x)) // Add the value 1 to inc(x)
```

Functions of a single argument are frequently referred to as *unary* functions, and unary functions whose return type is `boolean` are referred to as *predicates*. Functions of two arguments are frequently referred to as *binary* functions, while binary functions whose return type is `boolean` are referred to as *relations*.

Rosetta function application is lazy and defined using curried function semantics. Curried function semantics turns multi-parameter functions into a sequence of unary function applications. For example:

```
add(1,inc(x)) == (add(1))(inc(x))
```

The application of the two-argument `add` function to its first argument results in a unary function that adds 1 to its argument. When this result is applied to `inc(x)`, it adds one to the result of the function application. It is perfectly legal to partially specify function arguments. Specifically, the following definition of `inc` can be perfectly legal:

```
inc == add(1)
```

The `inc` function that increments its argument is equivalent to the add function with one argument instantiated with one. Currying and function evaluation will be explained later. For now one need only understand that function applications are applications of function items to formal parameters and that curried functions are allowed.

The type of a function application whose formal parameters are of the correct type is the return type specified in its declaration. If the formal parameters are not of the correct type, the function application's type cannot be determined and it is not soundly typed. Consider a simple function declaration for an increment function:

```
inc(x::natural)::natural is x+1;
```

Legal applications of `inc` include:

```
inc(1)
inc(inc(1))
inc(2+3)
```

In each case, the formal parameter to `inc` can be determined to be a subtype of `natural`. Thus, the type of each application is `natural`, the type specified by `inc`'s declaration. In contrast, illegal applications of `inc` include:

```
inc('a')
inc(-1)
inc([1,2,3])
```

In each case, the formal parameter is not a subtype of `natural`, no type can be inferred, and the application is not well-typed.

3.3 Operator Application

An *operator application* is a special syntax for function application. For example:

```
x + y    // add x to y
```

```
%x        // convert x from boolean to bit
1,2,3     // form a set containing 1,2, and 3
```

are all operator applications and represent the following function applications:

```
_ + _ (x,y)
% _ (a)
{_}{1,2,3}
```

where the function name is obtained by replacing arguments with underscores.

Operators come in four forms, *infix*, *prefix*, *postfix*, and *mixfix*. The most common operator form is the traditional infix notation used for traditional mathematical operators. Infix operators use the syntax:

```
x ∘ y
```

where ∘ is the operator being applied and x and y are arguments to the operator. Examples of infix operations include:

```
1 + sqrt(x)     // add 1 to square root of x
x or y          // boolean disjunction of x and y
x & "␣error"    // concatenate x with a literal string
```

Prefix operations are also quite common and include many mathematical operators such as negation and size. The syntax for prefix operations is:

```
∘ x
```

where ∘ is the prefix operation and x is its argument. Examples of prefix operators include:

```
-x  // negate x
#x  // size of x
%x  // convert x to a boolean or bit
```

Mixfix operators are special operators, where the arguments are mixed within the operation. There are several forms of mixfix operations, including *formers* and *if expressions*. Formers are used to collect items and form new items representing collections such as sequences and sets. The syntax of a former is generally:

```
< arglist >
```

where "<" and ">" represent the name of the former and arglist is a comma-separated collection of arguments. Examples of formers include set formers, sequence formers, and function formers:

```
{1,2,3}         // set containing 1,2, and 3
{*1,1,3*}       // multiset containing 1,1, and 3
[1,2,3]         // sequence containing 1,2, and 3
<* (x::integer)::integer *>        // function type or signature
<* (x::integer)::integer is x+1 *> // anonymous function
```

The syntax of the function former differs from the set and sequence formers.

Operators are special syntactic forms of function calls. In the case of pre- and postfix operations, the definition of the operation is provided by a unary function. In the case of infix operations, the defining function is binary. Mixfix operations differ based on the number of arguments, but still use a traditional function for their semantic definition. Special function names derived from operator symbols allow definition of operator functions. We take the operator and replace parameters with double underscores to form the name. For example, the names for the % and + operators are defined as follows:

```
%__    // function name for % operation
__+__  // function name for + operation
```

The following equivalences define mapping from operators to function applications:

$$\%x \stackrel{\text{def}}{=} \%\_\_(x)$$
$$a+b+c \stackrel{\text{def}}{=} \_\_+\_\_(\_\_+\_\_(a,b),c)$$

Now operators and functions share a common semantics for evaluation. Such equivalences, called *derived forms*, allow sharing of semantics across language constructs.

Because operators are special forms for functions, the types associated with operator application are defined in the same manner. Specifically, if operator arguments are of a type compatible with the operator, then the operator application type is obtained from the function signature. Unfortunately, overloading of operators complicates this process substantially, with the number operations introducing significant complexity. Chapter 4 discusses operators over basic types extensively and describes operator type inference.

3.4 If Expressions

An *if expression* specifies choice between values based on a boolean expression. The **if** expression:

```
if x>0.3 then 1 else 0 end if
```

equals 1 if x>0.3 and 0 if it does not. Thus the term:

```
x' = if x>0.3 then 1 else 0 end if
```

asserts that the next value of x equals the result of evaluating the **if** expression.

Most traditional programming languages use an *if statement* rather than an **if** expression. The distinction is that an **if** expression, like all expressions, has an associated value. An **if** statement simply orders the application of other statements. This distinction appears in the previous example. It can be rewritten in a form similar to an **if** statement by lifting the **if** outside the equality:

```
if x>0.3 then x'=1 else x'=0 end if
```

The net result is the same — an assertion of a value for x in the next state. The formal syntax of the **if** expression uses the traditional block style more typical of if statements and includes an **elseif** option:

```
if c₀ then e₀
  [elseif cₖ then eₖ]*
  else eₙ
end if
```

In the **if** expression, all c_j expressions must be of type **boolean** while all e_j expressions must be of the same type. This assures that the **if** expression will have a single type. The **if** expression evaluates to the first e_j associated with a true condition. If no conditions are true, then evaluation results in e_n associated with the **else** keyword. Although the **if** expression is block-oriented and reminiscent of imperative languages, it is a true expression and not a statement. Like all expressions, **if** expressions have values and types.

The most common form of the **if** expression is the simple if-then-else form:

```
if eₑ then eₜ else e_f end if;
```

The rule for evaluating this expression is a simplification of the general semantics. When e_c is true, the expression is equal to e_t. When e_c is false, the expression is equal to e_f. Specifically:

$$\text{if true then } e_t \text{ else } e_f \text{ end if} \stackrel{\text{def}}{=} e_t$$

$$\text{if true then } e_t \text{ else } e_f \text{ end if} \stackrel{\text{def}}{=} e_f$$

These two equivalences define two-thirds of the **if** semantics. The final rule states that when e_t is an expression other than a boolean value, it must be evaluated first.

Understanding the **elseif** construct is a simple extension of the **if** expression presented so far. The following expression:

```
x' = if x>0 then 1
        elseif x<0 then -1
        else 0
     end if
```

is equivalent to:

```
x' = if x>0 then 1
        else if x<0 then -1
                else 0
             end if;
     end if;
```

Applying the previous rules to this form results in the expected behavior. If x>0 the expression is equal to 1. If x<0 then the expression is equal to -1. Otherwise the expression is equal to 0.

The **if** expression must be viewed like all other operators as a function with an associated type and value. The value is determined by evaluating one of the two expressions based on a boolean condition. Again, the **if** expression does not define paths or sequences of instructions in a Rosetta specification.

EXAMPLE 3.1

If Expressions

Evaluating the following form:

```
if x=1 then f(x) else g(x) end if;
```

results in f(1) when x=1 and g(x) otherwise. Note that the expression is replaced by these results where it occurs in a specification.

Evaluating the following form:

```
if p(x) then f(x)
  elseif q(x) then g(x)
  else h(x)
end if;
```

results in f(x) when p(x) is true; g(x) when p(x) is false and q(x) is true; and h(x) when both p(x) and q(x) are false.

Each expression, f(x), g(x), and h(x), must have the same type. ■

The type of an **if** expression is derived from the sub-expressions defined in the **then** and **else** clauses. The conditional expression must be of type **boolean** and the **then** and **else** expressions must be of same type. If these conditions hold, the type of the **if** expression is the type of the **then** and **else** expressions. When **elseif** clauses are present, they must share the type of the **then** and **else** clauses. The following if expression defines a signumm, or sign, operation:

```
if x >= 0 then 1 else -1 end if;
```

x >= 0 is a boolean expression, thus the type of the **if** expression is the common type of 1 and -1, or integer. It can be used wherever an integer is allowed. The following forms are legal uses of this expression:

```
(if x >= 0 then 1 else -1 end if) + 1;

sgn(x::integer)::integer is if x>= 0 then 1 else -1 end if;
```

In contrast, the following forms are not legal uses:

```
s(if x >=0 then 1 else -1 end if)

(if x >=0 then 1 else -1 end if) and 1
```

In the first case, sequence access must use a natural value. Because the **if** expression is of type integer, there is no guarantee that when evaluated the result can be used to access a sequence element. In the second case, **and** expects bit arguments.

Although the **if** expression can produce a value 1::bit, it is not guaranteed to produce a bit.

EXAMPLE 3.2

If and Conditional
Equivalence

Among the most common uses of **if** statements in a traditional programming language is to provide a mechanism for conditional assignment. A simple case of such a description from the digital design domain is the definition of a multiplexer. A naive specification of a Rosetta multiplexer has the form:

```
facet mux(i0,i1,c::input bit; z::output bit)::state_based is
begin
  if c=0 then z'=i0 else z'=i1 end if;
end facet mux;
```

In this case, the **if** expression evaluates to either z'=i0 or z'=i1, depending on the value of c. To completely understand this specification, it is important to note that z' refers to the next value of z much like the (VHDL) after clause that specifies values for signals sometime in the future.

A more succinct method of writing the multiplexer **if** expression in Rosetta takes advantage of the **if** expression feature that it is in essence a function. By lifting the equivalence out of the **if** expression, the mux facet can be rewritten as:

```
facet mux(i0,i1,c::input bit; z::output
bit)::state_based is begin
  z'=if c=0 then i0 else i1 end if;
end facet mux;
```

Semantically, the two definitions are identical. However, the second specification more directly defines what the multiplexer does. If c=0 is true, then the **if** expression simplifies to i0. Otherwise, it simplifies to i1. ∎

It should be noted that if an **if** expression's type is determined to be **top**, it is virtually useless as a specification expression. If a system is to be well-typed, the only place such an expression can be used is where a **top** is expected. Few functions are general enough to operate on **top** typed arguments.

EXAMPLE 3.3

Type Inference for If
Expressions

Some examples of **if** expressions and associated types are shown below. Because we have not defined many of Rosetta's basic types, some intuition must be used to think about expression types. The negint type is the set of negative integers and is a subtype of integer. The bit type is the set {0,1} and is a subtype of natural, which is in turn a subtype of integer. The imaginary and real types are both subtypes of complex. These types will be defined later; for now, intuition should be sufficient to understand these typing examples.

In the expression:

```
if true then -1 else 1 end if :: integer
```

the condition is boolean and the clauses are both elements of integer. Even though the condition is always true and the expression reduces to -1, the inferred type is integer, not negint. Evaluation of the conditional expression is not and cannot be performed during type inference. If a static analysis tool can make this determinations, then a type assertion can be added to inform the type checking system of this fact.

In the following expression, where j is the imaginary root:

```
if x<1 then j else -1 end if :: complex
```

the condition is boolean, one expression is imaginary, and the other is negint. The common type for these types is complex and thus the expression type is complex.

In the expression:

```
if x=y then j else -j end if :: imaginary
```

the condition is boolean and both expressions are imaginary, so the expression type is imaginary.

In the expression:

```
if x>0 then 1
  elseif x=0 then 0
  else -1
end if :: integer
```

the condition is boolean and the expressions are typed bit, bit, and negint, respectively. The common type is integer, thus the if expression is type integer.

Finally, in the expression:

```
if x>0 then 1
  elseif x=0 then 'a'
  else -1
end if :: top
```

the condition is boolean and the expressions are bit, character, and negint, respectively. The only common type is top, so the if expression is type top. ∎

3.5 Case Expressions

A *case expression* allows selection from among many alternatives based on set membership. The **case** expression is much like the **if** expression, but checks to determine if the result of evaluating an expression is contained in a set of items, rather than checking a boolean value. For example:

```
case x is
{0,1,2} -> x+1 |
{3} -> 0
end case;
```

represents a state transition function for a modulo four counter. If the value of x is in the set {0,1,2}, then the case expression evaluates to x+1. If the value of x is in the set {3}, then the form evaluates to 0. Like the **if** expression, the **case** expression evaluates to a value. Assuming that x is the current state and x' is the next state, the following form defines a mechanism for calculating the next state in a zero to three counter:

```
x' = case x is
  {0,1,2} -> x+1 |
  {3} -> 0
end case;
```

The general syntax for a **case** expression is:

```
case e is
  s0 -> e0
  [ | sn -> en ]*
end case;
```

The **case** expression evaluates to e_k associated with the first s_k containing the result of evaluating e. If e is not contained in any s_k, then the case expression is **bottom**. No default case is provided; however, using the type **top** will result in a default **case**. Specifically, the **top** type is defined as the set containing all possible Rosetta values. Thus, any value of e will be an element of **top**. Specifically:

```
case x is
  {0,1,2} -> x+1 |
  {3} -> 0 |
  top -> bottom
end case;
```

will evaluate to bottom if x is not contained in either the set {0,1,2} or the set {3}. The value bottom is an element of all sets and will commonly be used to represent an error or default case when choices are specified. This case specifies the same behavior as leaving the **top** case off. An alternate approach specifies a kind of reset behavior when the selection expression is out of range:

```
case x is
  {0,1,2} -> x+1 |
  {3} -> 0 |
  top -> 0
end case;
```

In this example, the expression will evaluate to 0 when the selection expression is out of range. A counter using this definition will always reset to 0 when the current state is illegal.

The previous example illustrates several common Rosetta specification techniques that will be examined throughout the remainder of this text. The terms are declarative constructs that must be simultaneously true. Thus, the output value, the next state value, and the state type are constrained concurrently, not sequentially. Both the **case** and **if** expressions have values and are not sequencing statements. Finally, the use of the decorated variable, current', denotes the value of current in the next state.

Determining the type of a **case** expression is identical to determining the type of an **if** expression. The most common of the possible result expressions is the type of the **case** expression. Unlike the **if** expression, there is no requirement on the selection expression. The presence of the **top** type again assures that a least common supertype is available.

EXAMPLE 3.4

Case Expressions and State Machines

State-based specification is one of the most common mechanisms used for abstract system design. In state-based systems, a set of states is specified along with state transition and output functions defining next state and output respectively. The following facet defines requirements for a modulo-4 counter that outputs its state value.

```
facet counter(clk,reset::input bit;
              value::output word(2))::state_based is
  current::state;
begin
  state_def: state = word(2);
  next_state: current' = if event(clk) and clk=1
      then if reset=1 then b"00"
              else case current is
                  b"00" -> b"01" |
                  b"01" -> b"10" |
                  b"10" -> b"11" |
                  b"11" -> b"00"
              end case
          end if
      else current;
      end if;
  output_ val: value = current;
end facet counter;
```

In this definition, the case expression defines the next state function by listing all possibilities of state and their associated values. The facet specification defines a model with two binary inputs, clk representing the clock input, and **reset** representing a synchronous reset operation. The variable current is the value of the current state. The heart of the specification is three terms specifying the state type, state value and output functions. These terms are labeled state_def, next_state, and output_val respectively.

The state_def term defines the state type to be word(2), a 2-bit binary word. This allows the remainder of the specification to use the state value in expressions if necessary. The output_val term defines the values of the value output by equating it with the current state. This term defines an invariant and makes the machine a Moore machine because outputs are associated with state only.

The next_state term defines how state changes using **if** and case expressions. The outer **if** determines if an event has occurred on the clock and whether the resulting value is 1. This is a common mechanism for specifying the rising edge of a clock. If this condition fails, the value of the **if** expression is current and the system maintains its state. The inner **if** checks the status of the synchronous reset signal. If it is high, the value of the **if** expression is b"00", the Rosetta representation for 2-bit, binary zero. If the reset value is not high, the case expression specifies the value of the **if** expression. The value resulting from evaluating the **if** expressions constrains the value of current' representing the value of current in the next state.   ■

EXAMPLE 3.5

Type Inference for Case
Expressions

Assuming that x is a natural number, the type of the case statement previously defined is natural:

```
case x is
  0,1,2 -> x+1 |
  3 -> 0 |
  top -> 0
end case :: natural
```
■

3.6 Let Expressions

The *let expression* defines local items and their scopes. The expression:

```
let pi::real be 3.1415 in

    pi*radius*radius

end let;
```

evaluates to the area of a circle. The value of pi is defined to be the real value 3.1415 in the scope of the **let** expression.

The syntax of a **let** expression is:

$$\textbf{let}\ v_0 :: T_0\ \textbf{be}\ e_0\ [\![;\ v_n :: T_n\ \textbf{be}\ e_n\ ]\!]^*\ \textbf{in}$$

$$e$$

end let;

The **let** keyword is followed by a declarative section where one or more local items may be defined. The syntax of local item declarations is identical to the syntax of a traditional item declaration except that the keyword **is** is replaced by the keyword **be** for readability. Semantically, declarations are identical. Like traditional item declarations, **let** declarations are defined by specifying a name and type using the type membership operator. The **be** clause is required and specifies a value for the local item. Within the scope of the **let** expressions, these items behave as traditional items. Like traditional declarations, multiple items of the same type may be specified using a comma-separated collection of item names in the declaration. Within *e*, any variable defined in the **let** declaration section is visible. For example, given area::boolean:

```
let pi::real be 3.14159 in
  if area then pi*r^2 else 2*pi*r end if
end let;
```

defines a **let** expression that either calculates the area or radius of a circle. The **let** expression declares and defines a local value, pi, that is visible in the **if** expression used to calculate the indicated value.

Like all Rosetta expressions, the **let** expression simplifies to a value and has a type. In the previous example, the **let** expression's value is the **if** expression in the context of local variables. The **let** expression's type is the type of its encapsulated expression with the types of local variables added to the expression's context.

Let expressions may be nested in the traditional fashion. In the following specification, the variable x of type T_1 has the associated expression v_1, while y of type T_2 has the expression v_2. Both may be referenced in the expression e:

```
let x::T₁ be v₁ in
  let y::T₂ be v₂ in
    e
  end let
end let;
```

This expression may also be written as:

```
let x::T₁ be v₁; y::T₂ be v₂ in e end let;
```

Semantically, the **let** expression is actually a *letrec* expression. Languages such as Scheme and Common Lisp provide separate *let* and *letrec* constructs. A traditional **let** construct is not recursive. The defined variable cannot be referenced in the expression defining its value. A *letrec*, traditionally defined as a derived form of the traditional **let** construct, is recursive and allows such references.

EXAMPLE 3.6

Let Expressions and
Abstraction

A key use for the **let** expression is adding abstraction to definitions. Using the **let**, new declarations specific to an expression define and name concepts. In the following expression, a **let** expression defines quantities for use in a definition:

```
let newx::real be x+dx; newy::real be y+dy in
  update_position(newx,newy)
end let
```

Here the local variables newx and newy provide names for new values of an x and y position, respectively. Although this is a trivial use of the **let** expression, it demonstrates an abstraction capability that will prove more useful when recursive **let** applications and the fixed point operator, fix, are introduced.

The type of a **let** expression is the same as the type of its encapsulated expression. If the expression is well-typed, so is the associated **let** expression. For the previously defined **let** expression, the encapsulated expression is real, thus the **let** expression is real:

```
let (pi::real be 3.14159) in
  if area then pi*r^2 else 2*pi*r end if
end let:: real
```

3.7 Compound Expressions

Compound expressions are formed from atomic expressions, function applications, operator applications, and **if**, **case**, and **let** expressions by replacing formal parameters with expressions of an appropriate type. If e_k and g_k are legal expressions, then the following forms are examples of legal compound expressions:

```
e₁ ∘ e₂          // Infix operator application

∘ e₁             // Prefix operator application

(e₁)             // Parenthesized expression

f(e₁)            // Application of function f to e₁

[inc(e₁),3]      // Sequence of inc(e₁) and 3

if e₁            // If expression
  then e₂
  else e₃
end if

case e₁ of       // Case expression with 3 options
  {e₂} -> g₂ |
  {e₃} -> g₃ |
  top -> g_top    // Default case
end case

let v::T be e₁ in  // Let expression
    e₂
end let
```

Each of these compound expressions assumes that expressions are of a type compatible with their associated formal parameter or position in the expression. For example, the **if** expression mandates that e_1 be boolean and e_2 and e_3 share a common supertype. If type conditions are met, then these represent templates for legal compound expressions. Because e_k can be compound expressions, arbitrarily complex expressions can be formed through substitutions.

All Rosetta operations fall into one of several classes that define operator precedence, as illustrated in Table 3.1. Parenthesized expressions are of highest precedence followed by type assertions. Prefix, product, sum, and relational operations follow in that order. Function applications and mixfix value and type formers follow next, with equivalence having the lowest precedence. As other operations are defined, they will attempt to assume canonical positions in the precedence table. Parentheses may always be used to specify precedence explicitly rather than relying on precedence rules.

Table 3.1 Operator precedence table

Operation	Syntax
Parenthesized expression	(e)
Type assertions	e::T
Prefix operations	-e, #e, %e,...
Product operations	e * e, e / e,
	e^e, e **and** e
Sum operations	e + e, e - e,
	e **or** e, e => e,
	e <= e
Relational operations	e=e, e/=e,
	e<e, e=<e,
	e>e, e>=e
Function applications	f(e,e,e,...e)
Value and Type formers	[e,e,e,...],
	{e,e,e,...},
	{*e,e,e,...*},
	< *(v::T,v::T...)::T **is** e *>
Equivalence	e == e

When defined, all operation symbols are assigned to a class and remain in that class regardless of their functional definition. For example, $e_1 + e_2$ could be redefined to perform a multiplication operation. However, $e_1 + e_2$ will always have lower precedence as compared to a multiplication operation defined by $e_1 * e_2$. All binary operations are left associative unless otherwise specified.

Currently Rosetta provides no mechanism for defining precedence of new operations. If users introduce new operator syntax using domain structures, they are required to disambiguate precedence explicitly using parentheses. This is rarely an issue, as new operations are typically introduced using function definitions.

Elemental Types

Rosetta defines several type categories that include elemental, composite, function, constructed, and facet types. The first of these categories, the *elemental types*, defines types as collections of atomic values. Element literals provide a mechanism for specifying element values in a specification. All elemental types are subtypes of the element type and include numbers, boolean values and characters, infinite values, and the the **bottom** value.

The number type is the supertype of all definable number values. Subtypes of number include complex, real, integer, natural, and bit as well as specific subtypes to aid in constraining number values further for specification purposes. Number literals provide a mechanism for specifying number values in specifications.

The boolean type defines the two literal Boolean values **true** and **false** along with the traditional Boolean operations. Distinct from the bit type, the boolean type provides an abstract means for specifying logical sentences and for specifying conditional properties.

The character type defines a mechanism for specifying and manipulating Unicode values. Unicode is used throughout Rosetta to promote readable specifications and use of international character sets. The character literals provide a mechanism for defining Unicode values in a specification. The character type provides operations similar to those used in ASCII and Unicode character systems, allowing characters to be used in traditional ways.

4.1 The Boolean Type

Rosetta provides a Boolean type consisting of the values **true** and **false** denoted by the type boolean. The notation:

```
b::boolean;
```

Table 4.1 Operations defined over the `boolean` type

Operator	Syntax	Result Type
Logical negation	**not** A	`boolean`
Logical conjunction and disjunction	A **and** B, A **or** B	`boolean`
Logical nand and nor	A **nand** B,A **nor** B	`boolean`
Logic al exclusive or and nor	A **xor** B, A **xnor** B	`boolean`
Logical implication	A=>B,A **implies** B, B<=A	`boolean`
Boolean to bit	%A	

declares a new item, b, of type `boolean`. Operations provided over boolean provide a mechanism for specifying propositional statements in a traditional manner. Table 4.1 defines the collection of operations defined over all `boolean` items.

The unary **not** and binary **and** and **or** operators provide basic logical operations over `boolean` items. Examples of their application include:

```
true and true == true
false or false == false
false or true == true
false and true == false
not false == true
```

The **nand**, **nor**, **xor**, and **xnor** operators provide compound logical functions and are constructed from the basic logical operations. Although not semantically necessary, they provide shorthand for more complex, standard operations. Some examples of their application and definition include:

```
A nand B == not(A and B)
A nor B == not(A or B)
A xor B == ((not A) and B) or ((not B) and A)
A xnor B == (A and B) or ((not A) and (not B))
```

The various implication operators are also compound logical functions defined over disjunction. The logical implication operators, => and **implies**, are equivalent and define logical implication. The implied by operator, <=, is simply the inverse of the implies operation. Some examples of the application and definition of these operators are:

```
A implies B == not A or B
A => B == A implies B
B <= A == A implies B
```

The only operation on `boolean` that does not return a `boolean` is the boolean-to-bit operation. This operation facilitates moving easily between the two types. In Rosetta, the `boolean` and `bit` types are distinct, with `boolean` having no association with numbers and `bit` being a subtype of `natural` (defined later). Examples of application of this operation include:

```
%true == 1
%false == 0
%1 == true
%0 == false
```

In these examples, the boolean-to-bit operation is used as a bit-to-boolean operation. The same operator symbol is used for both conversions, allowing the definition:

```
%(%A) == A
```

When dealing with **bottom** argument values, operators over `boolean` are non-strict. If any argument to a `boolean` operator is the value **bottom**, but determining the value of the operator does not require evaluating **bottom**, then the result is the prescribed value. Otherwise, the result is **bottom**. Consider the following examples, where one argument is **bottom**, yet the value of the operator can be determined:

```
false and bottom == false
true or bottom == true
false implies bottom == true
```

In other cases, both operands must be known to determine a value other than **bottom** for the expression:

```
bottom and true == bottom
bottom or false == bottom
bottom xor false == bottom
bottom xor true == bottom
```

A good rule of thumb is that if the operator can be evaluated without evaluating **bottom**, then it will evaluate to the appropriate value.

4.2 The Number Types

Rosetta provides a wide collection of numeric types specifying sets of values commonly used in system specification. Within the number types, subtypes are arranged hierarchically, with each subtype inheriting operations from its supertype. The number type is the supertype of all number types. Each subtype adds new operations in addition to supporting operations from its supertype.

The `complex` type is the most general of the traditional number types and is formed from `real` and `imaginary` numbers in the classical fashion. The `real` type is further refined to define `posreal`, `negreal`, and `integer` types. The `integer` type is further refined to define posint, negint, and natural. Finally, natural is refined to define the bit types used in traditional digital design.

The only elements of the number type that are not also in complex are the infinite values, infinity and -infinity. These values are used primarily for specifying values for integrals, differential equations, and limits. However, they are in all ways number values.

4.2.1 Numbers

The type number is the supertype of all Rosetta numbers and a subtype of element. The number type is used heavily in the definition of other types, providing a collection of all numeric scalars. Specifiers use the number type to indicate that an item is simply a number. Given the declaration:

```
n :: number;
```

little can be inferred about the possible values of n except that they will be some number type.

The number type is useful when working on abstract specifications where an item type should be restricted to be a number, but with no other constraints. Although the complex type contains virtually all useful numbers, using complex can indicate that the specification in question must actually implement mechanisms for handling complex numbers.

4.2.2 Complex Numbers

Complex numbers are a direct subtype of number and are the most general number subtype. Denoted by the name complex, the set of complex numbers is constructed by adding real values to imaginary values in the same fashion as in traditional mathematics. The imaginary constant, j, is provided for this purpose. Expressions such as 3.124+4*j and inc(x)+sin(2x)*j result in complex values, as their syntax would suggest. Other mechanisms, such as exponentiation, that generate complex values are also allowed.

EXAMPLE 4.1

Using the Number Type

A *structural* specification describes a system by describing the components and interconnections that implement it. More specifically, a structural specification describes an architecture for a system. The following structural specification shows high-level architecture implementing serial composition of facet definitions:

```
facet comp[T::type](m::input T; n::output T)::state_based;

facet serial(x::input number;
             z::output number)::state_based is
  y::integer;
begin
  a: comp(x,y);
  b: comp(y,z);
end facet serial;
```

This specification defines a simple system, where component a accepts an input and communicates with component b, which generates an output. Note that all inputs and outputs are type number, but the interconnection item, y, is an integer. The facet comp is instantiated twice to represent the two components, but little is specified about comp other than its interface. Its interface uses polymorphic typing to allow any type to be communicated through the component, but the output type must be the same as the input type. When the components are instantiated with y, the type of their inputs and outputs is integer. Because the input to the component is specified as number, a more general type than integer, the serial component accepts a set of inputs larger than the component instantiating it. Thus, the specification will not type check properly. This problem is easily fixed by changing the interconnect variable type to integer or by making the serial architecture more general by allowing arbitrary types. ■

Operators defined over complex values are shown in Table 4.2 and include negation, sum and difference, multiplication and division, and exponentiation. The unary negation operator changes the sign of its argument while the unary identity operator returns its argument value. Examples of their application include:

```
-(3.0+4.0*j) == -3.0-4.0*j

+(3.0+4.0*j) == 3.0+4.0*j
```

The binary sum and difference operators are defined as traditional addition and subtraction of complex values. Examples of their application include:

```
(3.0+4.0*j) + (1.0-1.0*j) == 4.0+3.0*j

(3.0+4.0*j) - (1.0-1.0*j) == 2.0+5.0*j
```

Similarly, binary product and quotient operators are defined as traditional multiplication and division of complex numbers. Examples of their application include:

```
(2.0+2.0*j) * (1.0+4.0*j) == (2.0+10.0*j-8.0) == 6.0+10.0*j

(2.0+2.0*j) / (1.0+4.0*j) == (2.0+ 2.5*j-0.5) == 1.5+2.5*j
```

Because the sum and difference operators used in forming complex values are the same sum and difference operators used elsewhere, care must be taken to parenthesize complex values so that their formation takes higher precedence, compared to the operations performed on them. Removing parentheses in the previous examples has very different results.

The range of all operations defined on complex is also complex. This implies that operations are closed over complex and that their ranges cannot be restricted beyond complex. Functions, such as real projection or complex conjugate, can be restricted beyond complex and are presented subsequently.

In addition to unary and binary operators, a collection of predefined functions over complex are defined in Table 4.3. Unary functions for finding the real part,

Table 4.2 Operators over the complex type

Operation	Syntax	Result Type
Negation and Identity	-A, +A	complex
Sum and Difference	A+B, A-B	complex
Product and Quotient	A*B, A/B	complex
Exponentiation	A^B	complex

Table 4.3 Functions over the complex type

Function	Syntax	Result Type
Real and Imaginary parts	re(A), im(A)	real
Absolute value or modulus	abs(A)	real
Argument or phase	arg(A)	real
Complex conjugate	conj(A)	complex
Trigonometric functions	sin(x), cos(x), tan(x), arcsin(x), arccos(x), arctan(x)	complex
Hyperbolic trig functions	sinh(x), cosh(x), tanh(x), arcsinh(x), arccosh(x), arctanh(x)	complex
Exponential (base e)	exp(x)	complex
Square root	sqrt(x)	complex
Logarithms	log(x), log10(x), log2(x)	complex
Floor and ceiling	floor(x), ceiling(x)	complex
Truncation and rounding	trunc(x), round(x)	complex
Signum	sng(x)	complex

complex part, absolute value, argument, and complex conjugate all return real, a subtype of complex. The argument function's return value is in radians, as are all Rosetta functions dealing with angles. Some examples of the application of these functions include:

```
re(2.1-1.5*j) == 2.1

im(2.1-1.5*j) == -1.5

arg(2.1-1.5*j) == 0.9505

abs(2.1-1.5*j) == 2.587

conj(2.1-1.5*j) == 2.1+1.5*j
```

Functions for finding the floor, ceiling, and rounding complex values all return complex values determined in the traditional fashion. Examples of the application of these functions include:

```
floor(2.1-1.5*j) == 2.0-1.0*j

ceiling(2.1-1.5*j) == 3.0-2.0*j

trunc(2.1-1.5*j) == 2.0-1.0*j

round(2.1-1.5*j) == 2.0-2.0*j
```

The signum function over complex values returns a unit vector in the direction of the original value. In the following expression, sgn returns a vector of length 1 in the direction of the original vector:

```
sgn(2.1-1.5*j) == 0.8137-0.5812*j
```

A complete set of trigonometric operations is defined over complex numbers. The operations include the basic trigonometric functions for sine, cosine, and tangent as well as the hyperbolic trigonometric functions and inverse functions. Example applications include:

```
sin(2.0+3.0*j)

arcsin(2.0+3.0*j)

sinh(2.0+3.0*j)

arcsinh(2.0+3.0*j)
```

The sqrt function is provided as a shorthand for x^(1/2). The language definition constrains sqrt(x) value to one of the square roots of x. The function should be constrained further when specific roots are desired. For example, the following two equivalences both satisfy the base definition for sqrt:

```
sqrt(4.0) == 2.0

sqrt(4.0) == -2.0
```

The definition simply states that a square root function must return a legal root, not a specific legal root.

Similarly, a collection of logarithmic and exponentiation functions is defined that includes natural log, log base 10, log base 2, and exponentiation. Example applications include:

```
exp(2.0+3.0*j)

log(2.0+3.0*j)

log10(3.0+2.0*j)

log2(3.0+2.0*j)
```

Log base 2 is rarely, if ever, used for complex numbers, but is defined for completeness and will be useful for many subtypes of complex.

All trigonometric and logarithmic, functions are closed with respect to complex, the only exception being situations where operator forms are not allowed where their values are **bottom**, indicating an error. Such situations include division by 0 and tangent of 1.0. Some examples of illegal expressions whose value is **bottom** include:

```
3.0/0.0 == bottom

tan(0.0) == bottom

log(0.0) == bottom
```

Operators and functions defined for complex are strict. If any operand for a complex operator or function is **bottom**, then the result of evaluating that function is **bottom**. The implications of strictness are that if any operator is unknown or illegal, then the result of applying an operator or function is also undefined or illegal. This follows from the need to evaluate all arguments to mathematical operators and functions. This is in contrast to boolean operators that can often be evaluated without evaluating every argument.

4.2.3 Real and Imaginary Numbers

Both real and imaginary numbers are defined as subtypes of the complex numbers. Real numbers are the subtype of complex whose imaginary part is 0, while imaginary numbers are the subtype whose real part is 0. Real items are denoted using the type real, while imaginary numbers are denoted using the type imaginary and are typically formed by multiplying a real number by the imaginary root, j. Table 4.4 lists predefined constants associated with real and imaginary types. The imaginary root is used to construct imaginary values from real values. The standard mathematical constant, pi, has its canonical value, as does the exponential, e. Specifiers are free to design and package their own mathematical constants. However, these constants are provided by default in the base Rosetta modeling system.

Table 4.5 defines operations available on real and imaginary numbers in addition to those provided for complex numbers. The min and max operations return the minimum and maximum values from their arguments, respectively. Examples of their application include:

```
2.14 min 3.75 == 2.14

4.57*j min 2.76*j = 2.76*j

2.14 max 3.75 == 3.75

4.57*j max 2.76*j = 4.57*j
```

Table 4.4 Predefined numeric constants

Constant	Syntax	Type
Imaginary root	j	imaginary
Pi	pi	real
Exponential root	e	real

Table 4.5 Additional operations on real and imaginary types

Operator	Syntax	Result Type
Minimum and maximum	A **min** B, A **max** B	*Argument type*
Ordering relations	A<B, A=<B, A>=B, A>B	boolean

Min and max return types are the same as their associated argument types. If the arguments to either operation are imaginary, then the return type is imaginary, and similarly for real. Neither operator is defined for mixed imaginary and real values. For example, 4.75*j **max** 2.76 is forbidden.

Ordering relations provide traditional ordering relationships between values. Examples of their application include:

```
2.14 < 3.75 == true
2.14 > 3.75 == false
2.14*j =< 2.14*j == true
2.14*j >= 3.75*j == false
```

All ordering relations are boolean and, like the min and max operations, are undefined if their arguments are not both real or imaginary. For example, 2.14*j < 3.75 is not allowed.

Subtype relationships imply that any real or imaginary value may be treated like a complex value. Any operator defined over complex values may be applied to combinations of real and imaginary operands, with the result treated as the same type defined for complex arguments. However, for many operators and functions, the return type is more restricted than is the entire complex type. By specifying those restricted types, static analysis tools such as type checkers can infer more detailed information about a specification. Thus, Rosetta defines the return types for operators and functions as the most restricted type that covers all possibilities.

Sum and difference operators are closed with respect to both real and imaginary. The sum or difference of two real or imaginary values is real or imaginary, respectively. If operands are of mixed type, sum and difference both result in a complex type. This should not be surprising, as the sum and difference operations are used to construct the complex type from real and imaginary. For example:

```
1.2::real + 2.4::real == 3.6::real

(1.2*j)::imaginary + (2.4*j)::imaginary == (1.2*j)::imaginary

(1.2*j)::imaginary + (2.4)::real == (2.4+(3.6*j))::complex
```

Product operations are closed over `real`, but result in `real` when applied to `imaginary` and `imaginary` when applied to mixed arguments. Again, this should not be surprising, as j*j is equal to -1. Examples include:

```
1.2::real * 2.0::real == 2.4::real

(1.2*j)::imaginary * (2.0*j)::imaginary == -2.4::real

1.2::real * (2.0*j)::imaginary == (2.4*j)::imaginary
```

All operators and functions inherited from `complex` remain strict. All operators defined for `real` and `imaginary` are strict. If any of their operands or arguments are `bottom`, then they evaluate to `bottom`. This again results from the need to evaluate every argument to evaluate a mathematical operator.

4.2.4 Positive and Negative Numbers

The `posreal` and `negreal` types are subtypes of `real` that define the positive and negative real numbers, respectively. No additional operators are defined on either type, but the range of many operations on `real` changes when considering positive or negative real numbers.

The negation operator becomes a conversion operator between the types. The negation of a `posreal` results in a `negreal` and the negation of a `negreal` results in a `posreal`. For example:

```
-2.354::negreal

-(-2.354) == 2.354::posreal
```

The **min** and **max** operators can be further restricted. If either argument to the **max** operator is a `posreal`, then the result is a `posreal`, while if either argument to the **min** operator is a `negreal`, the result is a `negreal`. For example:

```
1.0::posreal min 4.0::posreal == 1.0::posreal

-1.0::negreal min 4.0::posreal == -1.0::negreal

-1.0::negreal min -4.0::negreal == -4.0::negreal
```

```
1.0::posreal max 4.0::posreal == 4.0::posreal

-1.0::negreal max 4.0::posreal == 4.0::real

-1.0::negreal max -4.0::negreal == -1.0::real
```

Sum and difference operators can both be thought of as sum operations by understanding the relation A-B == A+(-B). Both posreal and negreal are closed over addition. However, if sum is applied to mixed type operators, the result cannot be constrained beyond the real type. For example:

```
3.0::posreal + 2.0::posreal == 5.0::posreal

-3.0::negreal + -2.0::negreal == -5.0::negreal

-3.0::negreal + 2.0::posreal == -1.0::real

3.0::posreal + -2.0::negreal == 1.0::real
```

Product and quotient operations follow the same rules. Every division operation can be expressed as a product operation, thus we can define the return type for product and apply the result to quotient. Product is closed over posreal, while negreal results in a posreal. When operators are of mixed types, the result is always negreal. For example:

```
1.2::posreal * 2.0::posreal == 2.4::posreal

-1.2::negreal * 2.0::posreal == -2.4::negreal

-1.2::negreal * -2.0::negreal == 2.4::real
```

Operators over real inherited by posreal and negreal remain strict.

4.2.5 Integer Numbers

The integer type is the subtype of real that includes only integral values defined in the classical fashion. All operations defined over real are defined over integer. Three additional operations are listed in Table 4.6. The **div** operator provides an integer division operation, while **mod** and **rem** provide a modulus and remainder operation, respectively. The distinction between remainder and modulus is subtle, frequently causing the operations to be confused. The modulus operator specifies the

Table 4.6 New operations over integer type

Operator	Syntax	Return Type
Integer division	A **div** B	integer
Modulus	A **mod** B	integer
Remainder	A **rem** B	integer

integer modulus of its left operand divided by its right operand and assumes the sign of its right operand. The remainder operator specifies the integer remainder of its left operand divided by its right operand and assumes the sign of its left operand. For example:

```
11 div 3 == 3

11 rem 3 == 2

11 mod -3 == -2
```

All new operators and functions defined over integer are strict.

Integer operations inherited from real obey similar closure rules. Negation, sum, and product operators over integer are closed with the exception of division, which results in a real. For example:

```
2::integer * 3::integer == 6::integer

2::integer / 3::integer == (2/3)::real

2::integer + 3::integer == 5::integer

2::integer - 3::integer == -1::integer
```

Rounding operations such as round and trunc, floor and ceiling produce integer values for all real values and are thus closed over integer. However, with the exception of signum, such operators all reduce to the identity function.

The exponent operator and logarithmic operators are real valued when applied to integer. Although trigonometric functions are defined over integer as a subtype of real, they make little sense for discrete values.

4.2.6 Natural Numbers

Natural is the subtype of integer that includes only positive numbers and zero. The natural type inherits all operators and functions from the integer type, but introduces no new operations. Addition, multiplication, and exponentiation are both closed with respect to natural, as are **div**, **mod**, and **rem**. Subtraction and division are of type integer and real, respectively. For example:

```
1::natural + 2::natural == 3::natural

2::natural * 2::natural == 4::natural

2::natural ^ 3::natural == 8::natural

7::natural div 2::natural == 3::natural

7::natural mod 2::natural == 1::natural
```

```
1::natural - 2::natural == -1::integer

1::natural / 2::natural == (1/2)::real
```

Rounding and truncating functions are all closed over `natural`, as is the signum function. However, such functions are of limited use. Trigonometric and exponential functions are of type `real`.

4.2.7 Positive and Negative Integer Numbers

`Posint` and `negint` are subtypes of `natural` and `integer`, respectively. As their names imply, `posint` is the set of natural numbers without 0 and `negint` is the set of negative integers. Neither `posint` nor `negint` introduces new operations, but both inherit all operations from `integer`. All operators and functions over `natural` that are of `natural` type are of `posint` type when used over `posint`. Other operators and functions are of the same type when applied to `posint` as when applied to `integer`. Some examples include:

```
-(1)::posint == -1::negint

1::posint + 2::posint == 3::posint

1::posint - 2::posint == -1::integer

2::posint * 3::posint == 6::posint

2::posint / 3::posint == (2/3)::real

2::posint ^ 3::posint == 8::posint

2::posint max 3::posint == 3::posint

2::posint min 3::posint == 2::posint
```

Operators and functions from `integer` over `negint` present a more complicated problem. Addition is the only operator closed with respect to `negint`. Some examples include:

```
-(-1)::negint == 1::posint

-1::negint + -2::negint == -3::negint

-1::negint - -2::negint == 1::integer

-2::negint * -3::negint == 6::posint

-2::negint / -3::negint == (-2/-3)::real

-2::negint ^ -3::negint == -(1/8)::posint

-2::negint max -3::negint == -2::negint

-2::negint min -3::negint == -3::negint
```

Table 4.7 New functions and operations defined over the bit type

Operator	Syntax	Return Type
Logical negation	**not** A;	bit
Conjunction and disjunction	A **and** B, A **or** B	bit
Negated operators	A **nand** B, A **nor** B	bit
Exclusive or operators	A **xor** B, A **xnor** B	bit
Implication operators	A => B, A **implies** B, B <= A	bit

4.2.8 Bit Numbers

Bits are the subtype of natural numbers that includes only 1 and 0. Bit items are declared using the bit type and are used heavily in the specification of digital systems. Although all operators and functions defined on natural are also defined on bit, most specifiers use the Boolean functions introduced with the bit type. Table 4.7 lists new operators defined over the bit type. These operators parallel those specified for boolean; however, bit and boolean are distinct types. The %A conversion operator is provided to convert between these types. Examples of the bit operators include:

not 1 == 0

1 **and** 0 == 0

1 **or** 0 == 1

1 **xor** 1 == 0

1 **xnor** 0 == 0

%1 == **true**

%(%1) == 1

Like operations on boolean, operations defined for the bit type are non-strict. If arguments with values other than **bottom** are sufficient to determine the value of an operator, then the operator evaluates to its value. Examples include:

0 **and bottom** == 0

1 **or bottom** == 1

0 **implies bottom** == 1

Other operations inherited from natural are not closed over bit. Thus, any operation on bit using these operators and functions is treated as an operation on natural.

4.3 The Character Type

The character type is the collection of all UTF-32 values as defined in the Unicode Standard, Version 3.2 along with operations defined in Table 4.8. The ord and char functions move between the character values and the natural numbers. The ord function accepts a character value and specifies its associated Unicode value. The char function is its inverse, taking a Unicode value and specifying its associated Unicode value. For example:

```
char(ord(c::character)) == c::character

ord('a') == 16\0061\

ord('U+2132') == 16\2132\
```

The uc and dc functions change a character's case to upper or lower case, respectively, when such a conversion is defined. When the upper and lower case distinction does not exist, these operations are identity functions. For example:

```
uc('a') == 'A'

dc('A') == 'a'

uc('1') == '1'

dc('1') == '1'
```

Ordering functions defined over character are defined by mapping to their Unicode values and using ordering operators defined in natural. A character is less than another if its associated Unicode value is less than the other's Unicode value. For example, assuming that x and y are characters, x<y and x=<y are defined:

```
x < y == ord(x) < ord(y)

x =< y == ord(x) =< ord(y)
```

Table 4.8 Operators defined over the character type

Operator	Syntax	Return Type
Ordinal	ord(a)	natural
Character	char(n)	character
Relational Operations	a<b, a=<b, a>b, a>=b	boolean
Capitalization	uc(a), dc(a)	character
Unicode constant former	'U+XXXX'	character
Character constant former	'x'	character

Examples using ordering operations over character include:

```
'a' =< 'b'

'a' < 'b'

'b' >= 'b'

'b' > 'a'
```

All operations over character types are strict. If any operand to a character function is **bottom**, then it is equal to **bottom**.

It is important to note that the character and number types are distinct. No subtype relationship exists between these types and neither inherits operators or functions from the other. Both inherit operators from the element type and its supertypes. Thus, statements such as 'c'<5 are not well defined statically because there is no less than function defined over character and natural. If an ordering relationship is defined over **top**, that ordering relation will be used in the previous statement. No such ordering is defined in the base language.

4.4 The Element Type

Rosetta provides a supertype of all elemental types denoted as element. The type element consists of all values from boolean, number, and character and contains all predefined, atomic Rosetta values. The element type is rarely used in specifications, but plays an important role in the definition of other Rosetta types.

4.5 The Top and Bottom Types

Rosetta introduces two types, **top** and **bottom**, that represent the supertype and subtype of all other types, respectively. The type **top** consists of all Rosetta values and is a supertype of all other types. Formally, for any Rosetta type T:

```
T =< top
```

Because Rosetta is reflective, this implies that **top** contains not only traditional values such as numbers, characters, and sequences, but all possible types, functions, and facets. The following declaration defines a new x that is of type **top**:

```
x :: top
```

The **top** type should rarely appear in specifications due to its generality. Nothing can be said about x other than it must be a Rosetta value. Although this may seem advantageous at times, **top** has the effect of removing all type information from anything it is associated with. What **top** does provide is a mechanism for defining operations that apply to all Rosetta items, including Rosetta specifications themselves. If we define a function:

```
f(x::top)::top;
```

it is defined over all Rosetta items. However, the declaration says nothing about the function's resultant type. Any expression can be a legal formal parameter to f, but nothing is known about its return value. Anytime **top** is used in this way, all type information associated with the result of applying the function is lost. Care should be taken not to use **top** to try and define polymorphic functions.

The **bottom** type is a special type with a single value also called **bottom**; this is used to indicate error results and divergent computations. It allows errors to be specified without violating type conditions. **bottom** is the opposite of **top** in that instead of being a supertype of all types, it is a subtype of all types. Formally, for any Rosetta type T:

```
bottom =< T
bottom in T
```

bottom is inhabited by the single value **bottom** used to represent the value of a divergent computation. Like **top**, **bottom** should rarely be used in specifications. However, it does play an important role in defining Rosetta semantics. For example, the function example defines a function whose evaluation can result in **bottom**:

```
example(x::real)::real is
  if not(x==0) then f(x) else bottom end if;
```

If the input parameter x is ever 0, then the function is defined to be **bottom**, indicating an error condition. Because **bottom** is a subtype of every type, the **bottom** value always results in a type correct result.

4.6 Element Literals

Literals allow the user to write specific values in a Rosetta definition. Each elemental type has a mechanism for specifying literals. The number type provides a simple syntax to support intuitive specification of integer, real, and complex values. In addition, a rich syntax is provided to specify numbers with arbitrary radix and exponentiation capabilities. The character type provides a syntax for defining Unicode characters using character codes as well as simple literal characters for keyboard characters. The boolean type provides the Boolean literals **true** and **false** for specifying propositional statements. Finally, the bottom literal **bottom** provides a mechanism for denoting an undefined or error value.

4.6.1 **Number Literals**

Rosetta provides a mechanism for defining number literals in specifications that supports specifying numerical values ranging from simple decimal literals to exponential forms of numbers with arbitrary radix. The simplest number literals take the form of decimal digit strings with optional signs and a single decimal point. The following examples are all legal number literals of this form:

```
123     // Integer literal
12.3    // Real literal
-123    // Negative integer literal
-1.23   // Negative real literal
```

An exponent can be added to the literal value to specify the position of the radix point. The radix point position specifies the number of places the radix point should be moved to the left (negative point position) or right (positive point position). Expressing no radix position is the same as specifying the radix point as 0. The symbol e is used to separate the radix position from the remainder of the number specification. The following examples show this literal notation:

```
1.23e5  == 123,000       // 1.23*10⁵
1.23E-5 == 0.0000123     // 1.23*10⁻⁵
1.23e0  == 1.23          // 1.23*10⁰
```

The most general form for number literals adds a radix, allowing specification of numbers in arbitrary bases up to 16. The radix is added to the front of a number literal and separated from the value using a backslash ("\"). When a radix is added, the radix point position must also be separated from the value using a backslash. Using this notation, number literals are specified using the form:

```
radix \ mantissa \ e pointposition
```

When a radix is specified, the mantissa becomes a sequence of based digits rather than decimal digits. To support bases up to base 16, the decimal digits are extended to include A-F and a-f in the classical manner. The radix point position is always specified in decimal regardless of the radix, and is optional. The backslash separating the radix point position from the mantissa is never optional if a radix is specified. The value 16\234C\e4 is read "234C base 16 times 16 to the 4th power." Some examples include:

```
10\5.2\         // 5.2 in base 10
8\5.2\          // 5.2 in base 8
16\AABC\        // AABC in base 16
-10\5.2\e-10    // −5.2*10⁻¹⁰
10\5.2\e-10     // 5.2*10⁻¹⁰
16\5.2\e-10     // 5.2*16⁻ᴬ base 16
-2\1\           // −1 in binary
-2\1\e1         // −10 in binary
```

The backslash characters in a number literal are a part of its lexical structure and should not be viewed as operators: 10\5.2\e-10 is a number literal and should be viewed in the same manner as 5.2e-10. Thus, when a negation operator appears before literals of this form, it negates the entire value, not the radix value: -10\5.2\e-10 is a negative value in the same manner as -5.2e-10. The notation 10\-5.2\e-10 attempts to specify the negation as a part of the inner value, but is illegal in the notation. Although negations may appear in the radix point specification, the value must be unsigned.

In addition to values specified using the general form, three number constants are defined that can be used directly in specifications:

```
e      // Exponential

pi     // Geometric pi

j      // Imaginary root
```

A number literal may belong to several types. An extreme example of this is the literal 0 that belongs to every number subtype. When a number literal appears in a specification, it assumes the most specific type from those of which it is a member. For example, the following literals are asserted to be of the specified type:

```
0::bit

1::bit

2::posint

-2::integer

2.1::posreal

-2.1::negreal

(2.1+3.2*j)::complex
```

As we shall see later, this ensures that the appropriate type results when an operator is applied.

The literal infinity and its negation -infinity represent positive and negative infinite values. Although they are number values, they are not subtypes of any of the number subtypes. Any mathematical operation defined thus far applied to infinite values will result in the number type. Specifically:

```
1.3 + infinity == infinity::number

2 - infinity == -infinity::number
```

The infinite values are rarely used in this manner. The primary reason for their inclusion is with calculus functions such as limit and integral defined in the continuous time domain. There may be reasons to use infinite values in other kinds of specification, thus they are included in the number hierarchy.

4.6.2 **Character Literals**

To support specification across multiple languages and cultures, Rosetta uses the Unicode standard for character literals. The standard notation is `'U+XXXX'` or `'U-XXXX'`, where XXXX is a hexadecimal constant specifying the associated Unicode character. The tick marks are a part of the specification syntax and are always used when specifying character literals. Examples of Unicode character specification include:

```
'U+00B1'        // Plus/minus sign
'u+274f'        // Lower right drop--shadowed white square
'U+10347'       // Gothic letter IGGWS
'U+00FFFF'      // Not a character
'U-0001040F'    // Deseret capital letter yee
```

When the + symbol delineates the Unicode value, the short character code is used. When - is used, the full character code is specified.

Literal character values associated with keyboard symbols are specified using the notation `'x'`, where x is the literal character. For example, `'E'` is the literal E character, `'@'` is the "at" character, `' '` is the space character. These characters may be specified using the Unicode notation. This specification form is provided for convenience and readability.

Unlike number literals, `character` has no define subtypes. Thus, all `character` literals are simply of type `character`:

```
'a'::character
```

```
'U+274f'::character
```

```
'U-0001040F'::character
```

4.6.3 **Boolean Literals**

The only two literals of type `boolean` are the values **true** and **false**. Like `character`, `boolean` has no subtypes. Thus, all `boolean` literals are simply of type `boolean`:

```
true :: boolean
```

```
false :: boolean
```

4.6.4 **The Undefined Literal**

The undefined literal, called bottom and written as **bottom**, specifies a value that has no definition and is illegal. In this chapter, whenever evaluating an expression results in something that is undefined, its value is **bottom**. To facilitate using **bottom** as an error specifier, we say that **bottom** is an element of every type. Specifically, evaluating any expression, no matter what its type, may result in **bottom**. The undefined literal is rarely used by specifiers. However, the language definition uses it heavily to denote undefined or illegal calculations.

4.7 Operator Result Types

Operations on number, boolean, and character are defined with respect to operands and arguments of the same type. Addition, for example, is defined over types varying from bit through complex. By understanding the subtype hierarchy it is possible to determine specific operator instances for heterogeneous arguments.

Using the addition operator as an example, we can understand how Rosetta determines the specific operator instance to apply to any parameter pair. Knowing that, a type checker can determine the specific type associated with the addition operator application. Specifically, for any application $(A::T0+B::T1)::T2$, we can determine the type T2 by treating the arguments A and B as the most specific type where a matching operator instance exists. For example, the following applications of + demonstrate the approach:

```
0::bit + 2::natural == (0::natural + 2::natural)::natural

-1::negint + 0::bit == (-1::integer + 0::integer)::integer

1.1::posreal + 2::posint == (1.1::posreal + 2::posreal)::posreal

-1.1::negreal + 2.2::posreal == (-1.1::real + 2.2::real)::real
```

In each case, operands are treated as the most specific type where an instance of the addition operator exists. If such an instance cannot be found, then the expression is not properly typed and a type checker will indicate an error. The following examples show where an instance of addition cannot be found:

```
(0::bit + 'c'::character)

(false::boolean + true::boolean)

(true::boolean + 2.9::real)
```

In each case, no instance of the addition operator exists for the common supertypes of operand types. Bit and character share **top** as their only common supertype and no addition operator is defined there. The same holds for boolean and real as well as for boolean and boolean.

Composite Types

Rosetta's composite types define homogeneous collections of data. The composite type subsystem provides mechanisms for defining and manipulating three primary data containers, *sets*, *multisets*, and *sequences*. All composite types are homogeneous and are declared using *type formers*. Values from composite types are formed using associated *value formers*, whose syntax parallels type formers.

The *set* type defines unordered, homogeneous collections of unique items. A primary use for the *set* type is defining new types and subtypes. Additionally, sets provide excellent mechanisms for defining collections of items in abstract specifications. The *multiset* type defines an unordered, homogeneous collection of items, where multiple instances of an element are allowed. A primary use for *multiset* types is manipulating collections of values, where the number of values in a collection is significant. The *sequence* type defines indexed, homogeneous collections of data that resemble arrays and lists. A primary use for sequence types is ordering or structuring data collections. The sequence type has associated subtypes that represent strings and bit vectors as sequences of characters and bits, respectively.

5.1 Type Formers

All Rosetta composite types are defined using type formers. A type former is a function that takes one or more content types and generates a composite type with appropriate properties from them. All type formers have the following format:

container (T)

where *container* is the composite type and T is the type associated with items in the container. For example, in the definition:

```
bitset::set(bit);
```

set is the type former while bit is the content type. This declaration creates a new variable called bitset whose value is constrained to a set of values from the bit type. A constant is defined similarly:

```
mask::sequence(bit) is b"1100";
```

In this case, **sequence** is the type former while bit is again the content type. This declaration creates a new constant called mask whose type is a **sequence** of bit and whose constant value is the bitvector "1100".

The type former itself also defines a type. Excluding the content type from the previous declaration results in an item whose value is a set, but could be a set of elements from any arbitrary type:

```
anything::set
```

The use of **set** without the content type is the same as using **top** as the content type. Specifically, the preceding declaration is equivalent semantically to:

```
anything::set(top);
```

Type formers are simply Rosetta functions. Users can define their own type formers using techniques that will be documented in Chapter 8.

5.2 Set Types

A *set* is a unordered collection of unique items. The type former for set types is the keyword **set** and the form of a set declaration is:

```
s::set(T)
```

where s is the new item, and T is the type of elements contained in the set. Values associated with s may be any collection of items from T, including collections with zero elements. The distinction between this declaration and the declaration:

```
v::T;
```

is that in the second definition v must be a single element from T. For example, the definition:

```
x::set(integer)
```

defines a new item whose value can be any subset of integer. In contrast, the declaration:

```
v::integer
```

defines an item whose value is a single integer.

Of note is the type formed by using **top** as the type argument to **set**. The declaration **set(top)** defines a type that includes all possible sets of Rosetta values.

If an item has type **set(top)** we know that it is a set and nothing more. The set type former with no arguments denotes this type. Specifically, the following two declarations are semantically equivalent:

```
s::set(top) // Any set
```

```
s::set // Any set
```

Table 5.1 provides a list of operators defined over all types formed using the **set** type former. These operations include standard set theory operations as well as special-purpose formers for commonly used set values.

Not to be confused with the set type former, the set former is used to package expressions into sets. The set former uses the classical form for defining sets by extension. The notation {a,b,c} forms a set from the result of evaluating a, b, and c. For example:

```
{1,2,3}     // The set containing 1,2,3
```

```
{inc(1),2,3} // The set containing 2,3
```

```
{'a'}       // The singleton set containing 'a'
```

The set former is not a mechanism for defining set literals. It is syntactic sugar for a function that forms sets from values specified by arguments. The set former is strict, implying that if any of its arguments is **bottom**, then its value is **bottom**. This implies that no set can be formed with **bottom** as an element.

Table 5.1 Operators defined over sets

Operation	Syntax	Return Type
Set Former	{a,b,c,d}	set
Union, intersection, and difference	A+B, A*B, A-B	set
Set element	a in B	boolean
Subset and Superset	A=<B, B>=A	boolean
Proper Subset and Proper Superset	A<B, A>B	boolean
Cardinality	#A	natural
Contents	~A	set
Empty Set	{}	set
Integer Set	{i,..j}	set(integer)
Character Set	{c,..d}	set(character)
Powerset	set(A)	set(set(A))

The type of a set former application is the least common supertype of the elements specified as its arguments. For example:

{1,2,3}::**set**(posint)

{-1,0,1}::**set**(integer)

{inc(1.1)}::**set**(posreal)

{1,'1'}::**set**(**top**)

{{1,2},{2,3}}::**set**(**set**(posint))

The application of the set former with no arguments, {}, is defined as the set containing no elements, or simply the empty set. The empty set is a subset of all sets, no set is a subset of {}, and the size of {} is zero. Furthermore, {} unioned with any set is the original set and {} intersected with any set is {}. The type of the empty set presents interesting problems because it is a subset of all sets and thus an element of every formed set type. By convention, the empty set can be inferred to have any set type and this may be used as an operand to operators over sets involving any set type.

The element operator indicates when an item is an element of a set. The **in** operator is used to represent element and the notation a **in** S is read "a is an element of S." The relation is true when the item a is contained in the set S. For example:

3 in {1,2,3} == **true**

4 in {1,2,3} == **false**

The subset relation is true when all elements of its first set argument are contained in its second set argument. The =< and >= operators are used to represent subset operations. A=<B is read "A is a subset of B" while B>=A is read "B is a superset of A." Proper subset relationships hold when one set is a subset of another but is not equal to it. The the > and < operations provide proper containment. Thus, A<B is true if A=<B is true and A/=B. Likewise, A>B is true if A>=B and A/=B. For example:

{2,3} =< {1,2,3} == **true**

{2,3} < {1,2,3} == **true**

{1,2,3} < {1,2,3} == **false**

{1,2,3} >= {2,3} == **true**

{1,2,3} > {2,3} == **true**

{1,2,3} > {1,2,3} == **false**

Special operations are provided to form sets that contain sequences of integer and character values. The notation {i,..j} forms the set containing all integers between i and j inclusively if i and j are integer values. If i and j are character values, then a set of characters is formed. For example:

```
{2,..5}   == {2,3,4,5}

{'a',..'f'} == {'a','b','c','d','e','f'}
```

This notation is useful for numerous mapping operations, where sets of integer and character values must be generated.

Set union, intersection, and difference provide constructors for new sets from existing sets. The operator, +, generates the set union. Thus, A+B contains all elements contained in either A or B. For example:

```
{1,2,3} + {1,2,3} == {1,2,3}

{1,2,3} + {2,3} == {1,2,3}

{1,2,3} + {3,4,5,6} == {1,2,3,4,5,6}

{1,2,3} + {} == {1,2,3}
```

Intersection generates a new set containing all elements shared in its argument set. The operator, *, generates the set intersection. Thus, A*B generates a new set containing elements contained simultaneously in both set A and set B. For example:

```
{1,2,3} * {1,2,3} == {1,2,3}

{1,2,3} * {2,3} == {2,3}

{1,2,3} * {3,4,5,6} == {3}

{1,2,3} * {} == {}
```

Set difference generates a new set containing those elements in its first argument set that are not in its second argument set. The operator, −, generates the set difference. Thus, A−B generates a new set containing elements from A that are not in B. For example:

```
{1,2,3} − {1,2,3} == {}

{1,2,3} − {2,3} == {1}

{1,2,3} − {3,4,5,6} == {1,2}

{1,2,3} − {} == {1,2,3}
```

The cardinality operator, #, returns the size of a set. Thus, #A is the number of elements contained in A. For example:

```
#{1,2,3} == 3

#{} == 0
```

The contents operator, ~A, returns the contents of A as a set. For sets, the contents operator is an identity operator, as the elements of a set are the set itself. The contents operator is defined over all composite types and is included over set types for completeness.

The powerset operator, **set**, is used to generate collections of all possible subsets from a set. Thus, the notation **set**(A) generates the set of all possible subsets of A. For example:

set({1,2}) == {{},{1},{2},{1,2}}

It should be noted that the powerset operator is most often used for defining new types, because any Rosetta set can be used as a type. It is not a coincidence that the former for set types is the same as the powerset operation. For example, the declaration:

s :: **set**(integer);

asserts that s is an element of the powerset of integers. Because the elements of the powerset include all subsets of integers, the value of s is constrained to be one such subset. Thus, the set operator is used both to generate powersets and to define new set types. Because sets can be used as types in Rosetta, it is possible to construct user-defined types by providing functions that evaluate to sets. Examples of such usage can be found in Chapter 8.

The collection of functions defined over **set** types is defined in Table 5.2. The choose operator selects an arbitrary element from a set. Choose is not deterministic in that any element of its argument can be returned. For example:

choose({1,2,3}) == 1

choose({1,2,3}) == 2

The image function takes a unary function and a set and returns a new set resulting from the application of the unary operator to each element of the original set. For example, assuming that inc is defined as an increment function over integer:

image(inc,{1,2,3}) == {2,3,4}

The filter function takes a boolean predicate and a set and returns a new set that includes exactly those items from the original set that satisfy the predicate.

Table 5.2 Functions defined over sets

Function	Syntax	Return Type
Random choice	choose(s::A)	A
Function image	image(f,s)	ran(f)
Filter a set	filter(p,s::set(A))	set(A)

For example, assuming that `gtz` checks its argument to determine if it is greater than zero:

```
filter(gtz,{-1,0,1}) == {1}
```

`Filter` can be used to define new subtypes from types by filtering out values. The formal definition of the positive integer numbers is defined as the following set of values:

```
filter(gtz,natural) == posint
```

5.3 Multiset Types

A *multiset* is a collection of items that allows duplication of elements. Sometimes called bags, multisets differ from sets in that repeated elements are allowed and the number of element occurrences can be measured. The type former for multisets is the function **multiset**. The form of a **multiset** declaration is:

```
m::multiset(T)
```

where m is the new item and T is the type of elements contained in the multiset. Values associated with m may be any collection of items from T, including collections with zero elements. For example, the definition:

```
x::multiset(integer)
```

defines a new item whose value can be any **multiset** of integer values.

Of note is the type formed by using **top** as the type argument to **multiset**. The declaration **multiset(top)** defines a type that includes all possible multisets of Rosetta values. If an item has type **multiset(top)** we know that it is a multiset and nothing more. The **multiset** type former with no arguments denotes this type. Specifically, the following two declarations are semantically equivalent:

```
s::multiset(top) // Any multiset
```

```
s::multiset // Any multiset
```

Table 5.3 defines the collection of operations define over multisets.

The multiset former is used to package expressions into multisets and defines multisets by extension. The notation `{*a,b,c*}` forms a **multiset** from the result of evaluating a, b, and c. For example:

```
{* 1,2,1,3 *}        // The multiset containing 1,1,2,3

{* inc(1),2,1,3 *}   // The multiset containing 1,2,2,3

{* 'a' *}            // The singleton multiset containing 'a'
```

The multiset former is not a mechanism for defining multiset literals. It is syntactic sugar for functions that form multisets from values specified by arguments to the

Table 5.3 Operations defined over multisets

Operator	Syntax	Return Type
Multiset former	{∗a,b,a,d∗}	`multiset`
Union, intersection and difference	A+B, A∗B, A−B	`multiset`
Multiset element	a `in` M	`boolean`
Multisubset and multisuperset	A=<B, B>=A	`boolean`
Proper multisubset and proper multisuperset	A<B, B>A	`boolean`
Cardinality	#B	`natural`
Number in	x # M	`natural`
Contents	~A::`multiset`(A)	{`set`}(A)
Empty multiset	{∗∗}	`multiset`
Integer multiset	{∗i,..j∗}	`multiset`(integer)
Character multiset	{∗'c',..'d'∗}	`multiset`(character)
Power Multiset	`multiset`(A)	`set`(`multiset`(A))

former. The multiset former is strict, implying that if any of its arguments is `bottom`, then its value is **bottom**. This implies that no multiset can be formed with **bottom** as an element.

The type of a **multiset** former application is the least common supertype of the elements specified as its arguments. For example:

{∗ 1,2,1,3 ∗}::`multiset`(posint)

{∗ −1,−1,0,1,1 ∗}::`multiset`(integer)

{∗ inc(1.1) ∗}::`multiset`(posreal)

{∗ 1,'1' ∗}::`multiset`(top)

{∗{1,2,{1,2∗}::`multiset`(set(posint))

{∗{∗1,2∗},{∗1,2∗}∗}::`multiset`(**multiset**(posint))

The application of the multiset former with no arguments, {∗∗}, is defined as the multiset containing no elements, or simply the empty multiset. The empty multiset is a sub-multiset of all multisets, no multiset is a sub-multiset of {∗∗}, and the size of {∗∗} is zero. Furthermore, {∗∗} unioned with any multiset is the original multiset and {∗∗} intersected with any multiset is {∗∗}. The type of the empty multiset presents interesting problems because it is a sub-multiset of all multisets and thus an element of every formed multiset type. By convention, the empty multiset can be inferred to have any multiset type and may be used as an operand to operators over multisets involving any multiset type.

The element operator indicates when an item is an element of a multiset. The **in** operator is used to represent element and the notation a **in** M is read "a is an element of M." The relation is true when the item a is contained in the multiset M. For example:

3 **in** {∗1,2,3,3∗} == **true**

4 **in** {∗1,1,2,3∗} == **false**

The cardinality operator for multisets returns the number of elements in the multiset. The unary application of # to a multiset returns the cardinality of the multiset. Specifically:

#{∗1,1,2,1∗} == 4

#{∗∗} == 0

The cardinality operator may also be used as a binary operation that returns the number of occurrences of an element in a multiset. Specifically:

1 #{∗1,1,2,1∗} == 4

2 #{∗1,1,2,1∗} == 1

4 #{∗1,1,2,1∗} == 0

Using cardinality in this fashion is not defined for sets. If it were, the number of occurrences of any element in the set would be 1.

The contents operator defined on multisets returns the elements of the multiset as a set. The operation ~M takes an arbitrary multiset and returns its elements as a set, removing duplicate entries. For example:

~{∗1,1,2,1∗} == {1,2}

~{∗1,2,3,2,1∗} == {1,2,3}

~{∗∗} == {}

The sub-multiset relation is true when all elements of its first multiset argument are contained in its second multiset argument in at least as many occurrences as the first. The =< and >= operators are used to represent sub-multiset operations. A=<B is read "A is a sub-multiset of B," while B>=A is read "B is a super-multiset of A." Proper sub-multiset relationships hold when one multiset is a subset of another but is not equal to it. Two multisets are equal if they contain the same elements in the same quantities. The the > and < operations provide proper sub-multiset. Thus, A<B is true if A=<B is true and A/=B. Likewise, A>B is true if A>=B and A/=B. For example:

{∗2,3∗} =< {∗1,2,3,3∗} == **true**

{∗2,3,3∗} < {∗1,2,3∗} == **false**

```
{*2,3*} < {*y1,2,3*}  ==  true

{*1,2,3*} < {*1,2,3*}  ==  false

{*1,2,3,3*} >= {*2,3*}  ==  true

{*1,2,3*} > {*2,3,3*}  ==  false

{*1,2,3*} > {*2,3*}  ==  true

{*1,2,3*} > {*1,2,3*}  ==  false
```

Special operations are provided to form multisets that contain sequences of integer and character values. The notation {*i,..j*} forms the multiset containing one instance of all integers between i and j inclusively, if i and j are integer values. If i and j are character values, then a multiset of characters is formed. For example:

```
{*2,..5*}  ==  {*2,3,4,5*}

{*'a',..'f'*}  ==  {*'a','b','c','d','e','f'*}
```

Multiset union, intersection, and difference provide constructors for new multisets from existing multisets. The operator, +, generates the multiset union. Thus, A+B contains all elements contained in either A or B. Further, the number of any element in A+B is the sum of the numbers in A and B individually. If 'a'#A=1 and 'a'#B=2, then 'a'#(A+B)=3. For example:

```
{*1,2,3*} + {*1,2,3*}  ==  {*1,1,2,2,3,3*}

{*1,2,3*} + {*2,3*}  ==  {*1,2,2,3,3*}

{*1,2,3*} + {*3,4,5,6*}  ==  {*1,2,3,3,4,5,6*}

{*1,2,3*} + {**}  ==  {*1,2,3*}
```

Intersection generates a new multiset containing all elements shared in its argument multisets. The operator, *, generates the multiset intersection. Thus, A*B generates a new multiset containing elements contained simultaneously in both multiset A and multiset B. Further, the number of any element in the intersection is the minimum number in A and B individually. If 'a'#A==1 and 'a'#B==2, then 'a'#(A*B)=1. For example:

```
{*1,2,3*} * {*1,2,3*}  ==  {*1,2,3*}

{*1,2,2,3*} * {*2,3*}  ==  {*2,3*}

{*1,2,3*} * {*3,3,4,5,6 *}  ==  {*3*}

{*1,2,3*} * {**}  ==  {**}
```

Multiset difference generates a new multiset containing those elements in its first argument multiset that are not in its second argument multiset.

The operator, −, generates the multiset difference. Thus, A−B generates a new multiset containing elements from A that are not in B. Further, the number of any element in the difference is the difference between the number in the first multiset and the second. If 'a'#A=3 and 'a'#B=1, then 'a'#(A−B)=2. For example:

{*1,2,3*} − {*1,2,3*} == {**}

{*1,2,3*} − {*2,3*} == {*1*}

{*1,2,3,3*} − {*3,4*} == {*1,2,3*}

{*1,2,3,3*} − {*2,3,3*} = {*1*}

{*1,2,3*} − {**} == {*1,2,3*}

The power multiset operation, **multiset**(A), forms the set of all possible multisets formed from elements of its single **set** argument. It serves as the type former for multiset types. Unlike **set**(A), which is finite for all finite A, the value of **multiset**(A) is always infinite unless A is empty. This is due to multisets allowing arbitrary numbers of the same element. For example:

multiset({1}) == {{**},{*1*},{*1,1*},{*1,1,1*},...}

The definition of **multiset**(A) can also be defined as the set of multisets whose elements are taken from the set A. Formally, if a **in multiset**(A), then ~a =< A. The **multiset** operator is used almost exclusively for defining new multiset types, as it cannot be evaluated.

Functions defined over multisets are shown in Table 5.4. The choose operator selects an arbitrary element from a multiset. Choose is not deterministic in that any element of its argument can be returned. For example:

choose({*1,2,2,3*}) == 2

choose({*1,2,2,3*}) == 3

The image function takes a unary function and a multiset and returns a new multiset resulting from the application of the unary operator to each element of the original multiset. For example, assuming that inc is defined as an increment function over integer:

image(inc,{*1,2,2,3,3*}) == {*2,3,3,4,4*}

Table 5.4 Functions defined over multisets

Operator	Syntax	Return Type
Random choice	choose(m::A)	A
Image over multiset	image(f,m)	**multiset**(ran(f))
Filter over multiset	filter(p,m::**multiset**(A))	**multiset**(A)
Convert set to multiset	set2multiset(s::**set**(A))	**multiset**(A)

The `filter` function takes a Boolean predicate and a multiset and returns a new multiset that includes exactly those items from the original multiset that satisfy the predicate. For example, assuming that `gtz` checks its argument to determine if it is greater than zero:

filter(gtz,{*-1,-1,0,1,1,2,1*}) == {*1,1,2,1*}

Keep in mind that, like sets, the ordering of multiset elements has no bearing on its value. Two multisets are equivalent if they have the same contents. For example:

{*1,1,1,2,3,4*} == {*4,1,3,1,2,1*}

5.4 Sequence Types

A *sequence* is an indexed collection of items. Its fundamental use in the Rosetta system is defining ordered collections or lists of items. Sequences support random access and ordering of elements using their associated index, and in this regard behave much like arrays. Sequences also support concatenation allowing arbitrary creation of new sequences. The type former for sequences is the keyword **sequence**. The form of a sequence declaration is:

s::**sequence**(*T*);

where s is the new item, and *T* is the type of elements contained in the sequence. Legal elements of s are sequences of elements from *T*, including sequences with zero elements. For example, the definition:

x::**sequence**(integer)

defines a new item whose value can be any sequence of values from integer.

The declaration **sequence(top)** defines a type that includes all possible sequences of Rosetta values. If an item has type **sequence(top)** we know that it is a sequence and nothing more. The sequence type former with no arguments denotes this type. Specifically, the following two declarations are semantically equivalent:

s::**sequence(top)** // *Any sequence*

s::**sequence** // *Any sequence*

Table 5.5 provides a list of operators defined over all types formed using the **sequence** type former. Operators common to sequences as well as those defined for lists in other languages are included. In addition, operators for creating useful multiset values are defined.

The sequence former uses the classical form for defining sequences by extension. The notation [a,b,c] forms a sequence from the result of evaluating a, b, and c. The

Table 5.5 Operators defined over sequence types

Operator	Syntax	Return Type
Sequence Former	`[1,2,3,5]`	`sequence`
Concatenation	`s&t`	`sequence(A)`
Random access	`s::sequence(A)(n)`	`A`
Subscription	`s::sequence(A) sub i`	`sequence(A)`
Integer Sequence	`[i,..j]`	`sequence(integer)`
Character Sequence	`['c',..'d']`	`sequence(character)`
Subsequence relations	`s<t, s=<t, t>=s, t>s`	`boolean`
Minimum and maximum	`s min t, s max t`	`sequence`
Cardinality	`#s`	`natural`
Sequence Contents	`~s::sequence(A)`	`multiset(A)`
Empty sequence	`[]`	`sequence`

order of elements in the sequence former defines the order in the resulting sequence. The first element is associated with index 0 and other elements follow. For example:

```
[1,2,3]        // The sequence containing 1,2,3 in order

[inc(1),2,3]   // The sequence containing 2,2,3

['a']          // The sequence containing 'a'
```

The sequence former is not a mechanism for defining sequence literals. Like **set** and **multiset** formers, it is syntactic sugar for functions that form sequences from values specified by arguments to the former. The sequence former is strict, implying that if any of its arguments is **bottom**, then its value is **bottom**. This implies that no sequence can be formed with **bottom** as an element.

The type of a sequence former application is a sequence of the least common super-type of the elements specified as arguments. For example:

```
[1,2,3]::sequence(natural)

[-1,0,1]::sequence(integer)

[inc(1.1)]::sequence(posreal)

[1,'1']::sequence(top)

[{1,2},{2,3}]::sequence(set(posint))
```

The application of the sequence former with no arguments, `[]`, is defined as the sequence containing no elements, or simply the empty sequence. The empty sequence shares membership properties with the empty set and the empty multiset. The type of the empty sequence can be inferred to have any sequence type and may be used as an operand to operators over sets involving any set type.

Individual elements of a sequence type are accessed using their associated positional index. The sequence is used as a function name and the single argument is used to access an element. The notation s(2) accesses the element associated with index 2 in the sequence s; s(0) references the first element of the sequence. For example, assuming that s is equal to the sequence [3,3,2,1]:

    s(0) == 3

    s(2) == 2

    s(4) == **bottom**

As the example demonstrates, accessing a sequence element outside the known size of the sequence results in **bottom**.

It is not necessary to name the sequence to access its elements. The label, s, can be replaced by its value in the preceding example with equivalent results:

    [3,3,2,1](0) == 3

    [3,3,2,1](2) == 2

    [3,3,2,1](4) == **bottom**

Sequences of sequences are allowed and referencing is achieved by multiple levels of accessor functions. For example, s(0)(1) accesses the second element of the first sequence in s; s(0) references the first element of s. If s is a sequence of sequences, then s(0) is itself a sequence, thus s(0)(1) accesses an element in the sequence. Sequences can be arbitrarily nested in this way to provide functionality similar to a multidimensional array. For example, assume that we would like to define a two-dimensional array of integer values. The declaration for an example of this structure is:

    intarray :: **sequence**(**sequence**(integer));

Assume that intarray=[[1,2,3],[4,5,6],[7,8]]:

    intarray(0) == [1,2,3]

    intarray(1) == [4,5,6]

    intarray(1)(0) == [1,2,3](0) == 1

    intarray(2)(2) == [7,8](2) == **bottom**

The contents function, ~s, returns the contents of a sequence as a multiset. Thus the number of each element in the returned multiset is equal to the number of element occurrences in the sequence.

    ~[1,2,3,2,1] == {*1,1,2,2,3*}

    ~[1,1,1] == {*1,1,1*}

    ~[] == {**}

Because all individual sequences are finite, the contents multiset can always be formed.

The contents operator can be applied twice to obtain the set of values appearing in a sequence. The first application results in a multiset and the second results in a set. For example:

~(~[1,2,3,2,1]) == ~{*1,1,2,2,3*} == {1,2,3}

~(~[1,1,1]) == ~{*1,1,1*} == {1}

~(~[]) == ~{**} == {}

As with set and multiset, the size of a sequence type can be found using the size operator, #s; #s returns the natural number associated with the size of s. For example:

#[1,2,3,2,1] == 5

#[1,1,1] == 3

#[] == 0

The concatenation operator s&t is used to concatenate two sequences of the same type. Sequences are homogeneous data structures, thus all elements contained within them must be of a common type. The notation s&t produces a new sequence, with the elements of s preceding the elements of t. For example:

[1,2,3]&[4,5] == [1,2,3,4,5]

[1,2,3]&[] == [1,2,3]

[]&[1,2,3] == [1,2,3]

EXAMPLE 5.1

Sequence Forming and Concatenation

Assume the following declarations:

S::**sequence**(integer) is [1,2,3];

T::**sequence**(integer) is [1,2];

The following relationships hold:

```
S&T == [1,2,3,1,2];
T&S == [1,2,1,2,3];
S&S == [1,2,3,1,2,3];

S(0) == 1;
S(#S-1) == 3;
(S&T)(0) == S(0) == 1;
(S&T)(#S-1) == S(#S-1) == 3;
(S&T)(#S) == 1;
(S&T)(#(S&T)-1) == T(#T-1) == 2;

~S == {1,2,3};
~T == {1,2};
```

Assume the following declaration of a multidimensional sequence:

```
S::sequence(sequence(integer)) is [[1,2,3],[4,5,6]];
```

The following relationships hold:

```
S(0) == [1,2,3];
S(0)(1) == 2

S(1) == [4,5,6];
S(1)(0) == 4;

~S =={*[1,2,3],[4,5,6]*};
```

Note that concatenating a sequence of integers with the sequence S is not legal. To add a new sequence to the end of S requires concatenating with a sequence of integer sequences of the form [[5,6]]. Note that length is not an issue. Bounded sequences will be defined in a later section.   ∎

It is possible to extract a collection of elements from a sequence using an operation called subscription. If s is a sequence and i is a sequence of natural numbers, then **sub** i extracts the elements of s associated with values from i, generating a new sequence from those elements in the order specified by i. For example:

```
[1,2,3,4] sub [0,1] == [1,2]

[1,2,3,4] sub [1,0] == [2,1]

[1,2,3,4] sub [1,1,0,0,3,3,2,2] == [2,2,1,1,4,4,3,3]
```

Special operations are provided to form sequences of ordered integer and character values. The notation [i,..j] forms the sequence containing all integers in order between i and j inclusively, if i and j are integer values. If i and j are character values, then a sequence of characters is formed. For example:

```
[2,..5] == [2,3,4,5]

['a',..'f'] == ['a','b','c','d','e','f']
```

The =< and >= operators are used to represent subsequence relations between sequences. A=<B holds when A is a subsequence of B and A>=B when B is a subsequence of A. Similarly, the > and < operations provide strict lexicographical ordering relationships. A<B is true if A=<B is true and A/=B. Likewise, A>B is true if A>=B and A/=B. For example:

```
[1,2] =< [1,2,3,3] == true

[1,2,3] < [1,2,3] == false

[2,3] < [1,2,3] == true

[1,2,3] =< [1,2,3] == true
```

Sequence max and min operations are based on the subsequence relationships just defined. A **max** B returns A if A>B and B otherwise. Similarly, A **min** B returns A if A<B and B otherwise. Both operators can be defined using the **if** expression:

```
a min b == if a < b then a else b
```

```
a max b == if a < b then b else a
```

where a and b are both sequences. Table 5.6 lists functions available for sequence types.

Assume the following declarations of integer sequences:

```
s1::sequence(integer) is [1,2,3];
s2::sequence(integer) is [1,2,3,4];
```

Subsequence definitions state that the following relationships must hold:

```
s1 < s2;
s1 =< s2;
s1 =< s1;

s1 = s2 sub [0,..2];
s1 max s2 == s2 == [1,2,3,4];
s1 min s2 == s1 == [1,2,3];
```

Three operations are defined that allow sequences to be treated as lists. The accessors head and tail return the first element of a sequence and the remaining elements, respectively. The constructor cons adds a single element to the front of a sequence. Together, cons, head, and tail provide a list capability similar to that defined in traditional functional languages. The last function is also available to access the last element of a sequence. The head, tail and last functions are defined only on non-empty lists. Examples of their application include:

```
head([1,2,3]) == 1
```

```
tail([1,2,3]) == [2,3]
```

Table 5.6 Functions defined over sequence types

Function	Syntax	Return Type
Power Sequence	**sequence**(T)	**set(sequence**(T))
List accessors	head(s),	element,
	tail(s),	**sequence**,
	last(s)	element
Add to front	cons(e,s)	**sequence**
Reverse a sequence	reverse(s)	**sequence**
Image of function	image(F,s)	**sequence**
Comprehension	filter(P,s)	**sequence**
Reduce and Reduce tail	reduce(F,i,s),	element **of** the range **of** F
	reduce_tail(F,i,s)	
Element replacement	replace(s,n,v)	**sequence**
Zip	zip(F,s1,s2)	**sequence**

```
cons(1,[1,2,3]) == [1,1,2,3]

tail(cons(1,[1,2,3])) == [1,2,3]

head(cons(1,[1,2,3])) == 1

last([1,2,3]) == 3

cons(1,[]) == [1]

head([]) == bottom

tail([]) == bottom

last([]) == bottom
```

EXAMPLE 5.3

Head, Tail, and Cons Using
Subscription and
Concatenation

The functions head, tail, and cons can be defined using subscription, integer sequence, and the sequence former. The head of a sequence can be obtained by using the indexing operation to extract the first element. Formally, the head of a sequence can be defined as:

```
head(s) == s(0)
```

The tail of a sequence can be obtained by extracting all but the first element. This is a bit trickier, but the integer sequence operation can be used with subscription to extract all but the first element. The integer sequence operation generates the sequence 1 through length of S minus 1. Subscription is then used to extract all but the first element of the sequence:

```
tail(S) == s sub [1,..(#s-1)];
```

The cons constructor can be defined using the sequence former and concatenation to add an element to the beginning of a sequence:

```
cons(e,s) == [e]&s;
```

The sequence former creates a new sequence from e that is one element long. It is then added to the front of the sequence using the concatenation operator.   ■

The reverse function is defined over all sequences and evaluates to a sequence that contains the same elements as s, but in the reverse order. For example:

```
reverse([1,2,3]) == [3,2,1]

reverse([1]) == [1]

reverse([]) == []
```

The replace function generates a new sequence by replacing the element at position i with v. If replace attempts to access a sequence position beyond the length of s, the result is **bottom**. Unlike array operations in traditional languages, the original sequence is not modified in any way. A new sequence is generated whose elements match those in s except for position i contains the new value. For example:

```
replace(['a','b','c'],0,'b') == ['b','b','c']

replace(['a','b','c'],4,'b') == bottom
```

EXAMPLE **5.4**

Updating a Sequence

Sequences, like all other items, can be used as variables rather than as constant values. Updating sequences is quite easy using the replacement operation, as in the following statement:

```
seq' = replace(seq,2,5)
```

The result of evaluating this expression is constraining the value of seq in the next state, seq', to be the updated sequence from the current state — specifically, seq with the element indexed by 2 replaced with 5. Using the tick notation in this fashion will be fully explained in Chapter 12. For now, it is enough to simply realize that updating sequences like arrays can be achieved using replacement.   ■

The image and filter functions map functions onto sequence elements. The image function returns the sequence formed by applying function to each element of a sequence. If any application results in **bottom**, then image results in **bottom**. Specifically, if inc is a function that increments its argument by 1:

```
image(inc,[1,1,2,3]) == [inc(1),inc(1),inc(2),inc(3)] ==  [2,2,3,4]

image(inc,[]) == []
```

The filter function behaves much like the image function except that only the elements that a predicate holds true for are kept in the resultant list. Should any predicate application result in **bottom**, then the application of filter results in **bottom**. Specifically, if gtz is a function over integers that is true, if its argument is greater than zero:

```
filter(gtz,[-2,-1,0,1,2]) == [1,2]

filter(gtz,[]) == []
```

The reduce operator is a fold over sequences. It has two distinct forms that allow both left (reduce) and right (reduce_tail) associativity. Both forms take a sequence, a binary function whose domain is the same type as the sequence elements, and an initial value of the same type as the function range. The function is first applied to the initial value and the first sequence element. The function is then applied to each sequence element using the result of the previous application as the other argument.

A simple application of reduce defines the standard summation operator:

```
summation(s::sequence(integer))::integer is
    reduce(__+__,0,s);
```

The newly defined function will start with 0 and subsequently sum sequence values from left to right.

The `reduce_tail` function is similar except that the function is first applied to the initial value and the last element of the sequence, working from right to left. For operations like `summation`, where the applied function is commutative, the operators generate the same result. For example:

reduce(__+__,0,[1,2,3]) == (((0+1)+2)+3)

reduce_tail(__+__,0,[1,2,3]) == (1+(2+(3+0)))

In this case, the results are the same. However, for non-commutative operators like subtraction, the distinction is clear:

reduce(__-__,0,[1,2,3]) == (((0-1)-2)-3) == -6

reduce_tail(__-__,0,[1,2,3]) == (1-(2-(3-0))) == 2

Both forms are useful, but care must be taken when using them in definitions.

The `zip` is much like the `image` function except that it accepts two sequences an a binary function. The `zip` returns a list of elements resulting from the pair-wise application of the argument function to elements in the same position in the two lists.

If any function application results in **bottom**, then the application of `zip` also results in **bottom**. For example:

zip(__+__,[1,2,3],[0,1,2]) == [1+0,2+1,3+2] == [1,3,5]

zip(__/__,[1,2,3],[0,1,2]) == [1/0,2/1,3/2] == **bottom**

The power sequence operation, **sequence**(A), forms the set of all possible sequences formed from elements of its single `set` argument. It serves as the type former for sequence types. Like **multiset**(A), **sequence**(A) evaluates to an infinite set unless A is empty:

sequence(1) == {[],[1],[1,1],[1,1,1],...}

The definition of **sequences**(A) can also be defined as the set of sequences whose elements are taken from the set A. Formally, if a in **sequence**(A) then ~(~a) =< A. Like the **multiset** operator, the **sequence** operator is used almost exclusively for defining sequence types.

5.4.1 The String Type

A special case of sequence is the string type. Formally, string is defined as:

string == **sequence**(character);

The following notation defines a variable of type string:

str :: string;

A shorthand for forming constant strings is the classical notation embedding a sequence of characters in quotations. Specifically, "ABcdEF" is equivalent to the sequence ['A','B','c','d','E','F']. Thus, a string constant such as "$" can be defined as:

```
dollarString :: string is "$";
```

All operations that apply to sequences also apply to strings because strings are simply character sequences. In particular, the notation "abc"&"def" is useful for concatenation of strings. Ordering operations for sequences provide lexicographical ordering for strings based on the ordering of Unicode characters. Relationships like "ab"<"abc" are true as a result of subsequence operations.

Rosetta's string type provides operations that are typical of other similar languages. By treating strings as sequences of characters, operations are provided with little additional semantics. The following examples represent some useful operations over strings:

```
allcaps(s::string)::string is map uc s;

firstn(s::string, n::natural)::string is
    s sub [0..,n-1];

pal(s::string)::boolean is s == reverse(s);

contains(s::string, c::set(character))::boolean is

    (filter <*(x::character)::boolean is not(x in c)*> s) == [];
```

■

5.4.2 The Bitvector and Word Types

A second special case of sequence is the bitvector type used heavily in digital systems design. Formally, bitvector is defined:

```
bitvector == sequence(bit);
```

Thus, a bitvector is an array that contains only elements 0 and 1.

Bitvector literals may be specified either as binary or hexadecimal strings preceded by a the bitvector literal indicator b for binary or x for hexadecimal. The following bitvector literals define the same value:

```
b"10001100"
```

```
x"8C"
```

Operations over bit are generalized to bitvectors by performing each operation on similarly indexed bits from the two bitvectors using the zip operation. One such example of this is the definition of bitvector **and**. Assuming that b0 and b1 are equally sized bitvectors:

```
b0 and b1 = zip(__and__,b0,b1)
```

When applying bit operations to bitvectors, it is mandatory that the arguments be of the same length. The padding operators can easily be used to assure

Table 5.7 Operations defined on bitvectors in addition to traditional sequence operations

Operation	Syntax	Return Type
Binary operations	A **and** B, A **or** B, A **nand** B, A **nor** B, A **xor** B, A **xnor** B, **not** A	bitvector
Conversion to natural	bv2nat(b)	natural
Conversion from natural	nat2bv(n)	bitvector
Complement	twos(A)	
Shift operations	lshr(A), lshl(A), ashr(A), ashl(A)	bitvector bitvector
Rotate operations	rotr(A), rotl(A)	bitvector
Pad operations	padr(A,1,n), padl(A,1,n), sext(A,1)	bitvector

this requirement is satisfied. Assuming that b0 and b1 are bitvectors of unknown length less than 16 bits, they can be treated as 16-bit values as follows:

```
padl(b0,16,0) and padl(b1,16,0)
```

This use of padding assures that both operands have 16 bits. As we shall see later, Rosetta provides general padding and sign extension operators.

The operations bv2nat and nat2bv provide standard mechanisms for converting between binary and natural numbers. The binary value is treated as an unsigned value and converted in the canonical fashion. For example:

```
bv2nat(b"10") == 2
```

```
bv2nat(b"101") == 5
```

```
nat2bv(7) == "111"
```

```
nat2bv(0) == b"0"
```

It is always true that bv2n(n2bv(x)) == x.

The operation twos(bv) takes the two's complement of a binary value by negating and adding 1 to the result. This operator provides a negation function for fixed-length binary values. For example:

```
twos(b"011") == b"101"
```

```
twos(b"101") == b"011"
```

It is always true that twos(twos(x)) == x.

Two's complement operators operate on bitvectors with known lengths. To accommodate this, the twos function assumes that the resulting bitvector and the argument bitvector have the same length. The pad function can be used to appropriately size bitvectors prior to taking their complement, while the sign extend operator can be used to size the bitvectors after taking their complement.

Shifting operations are provided for manipulation of bitvectors. The lshr and lshl operations provide logical shift right and left, while ashr and ashl provide logical and arithmetic shifts right and left, the distinction being that logical shift operations shift in 0s while arithmetic shift operations shift in 1. The rotr and rotl operations provide rotation or circular shift. Examples include:

```
lshr(b"1100") == b"0110"

ashl(b"0011") == b"0111"

rotr(b"0011") == b"1001"

rotl(b"1011") == b"0111"
```

The padr(A,l,n) and padl(A,l,n) operations pad or concatenate a bitvector. Both functions take three arguments, a bitvector, a length value of type natural, and a pad value of type bit. If the length value is less than the length of the input bitvector, padr removes bits to the right and padl removes bits to the left, resulting in a vector of the specified length. In this case, the pad value is ignored. If the length value is greater than the length of the input bitvector, padr adds enough copies of the pad value to the right of the vector so that the return value is of the specified length. The padl operates similarly except that bits are added to the left of the bitvector argument. Some examples include:

```
padr(b"1100",2,0) == b"11"

padl(b"1100",2,0) == b"00"

padr(b"1100",6,1) == b"110011"

padl(b"1100",6,0) == b"001100"
```

The sext operation is a special case of padl and performs a sign extension on its argument. Sign extension is useful when padding a two's complement value without losing its sign. The operand is extended to the left using the value of its most significant bit rather than an explicit parameter. Some examples include:

```
sext(b"1100",2) == b"00"

sext(b"1100",6) == b"111100"

sext(b"0011",6) == b"000011"
```

Digital systems rarely deal with arbitrary bitvectors, but instead process fixed-length bitvectors. The type word is provided to define such fixed-length bitvector types. To define a word, the function word(n) is used, where n is the word length. For example:

```
w::word(8);
```

defines a variable w that can take the value of any bitvector of length 8. Neither shorter nor longer bitvectors belong to this type. Because word is a subtype of bitvector, all bitvector operations are defined on words.

It should be noted that bitvector and word types are not the same as binary numbers. Thus, the expression:

```
w = 2\10001001\;
```

is not legal because the argument types do not match. The type of w is word(8) while the type of 2\10001001\ is posint. Thus, the equivalence cannot be defined. The expressions:

```
w = nat2bv(2\10001001\);
```

```
w = nat2bv(16\89\);
```

are legal because in each case the number literal is converted to a bitvector prior to equating it with w.

EXAMPLE **5.6**

Slicing Bitvectors

An excellent example of subscription use is decoding an instruction in a CPU. Assume that I is a sequence of bits representing a typical 16-bit instruction. It is possible to decode I by extracting sub-sequence associated with instruction fields. Assuming that the instruction format is an instruction ID followed by three 4-bit register IDs, the following functions decode the instruction into its constituent parts:

```
op = I sub [15,..12];
```

```
rs = I sub [11,..8];
```

```
rt = I sub [7,..4];
```

```
rd = I sub [3,..0];
```

The value of rd may be used as a two's complement offset value using sign extend and the pad operation:

```
offset = sext(padr(rd,5,0),16)
```

In this case, a word address is generated by padding with 0 to the right. The sign extend function is then used to extend the 5-bit offset value to 16 bits without losing sign information.   ∎

6 Functions

Rosetta's function definition and evaluation capability provides function definition and application capabilities and a collection of advanced capabilities for more sophisticated and powerful specifications. Like functions in traditional functional programming languages, Rosetta functions provide a mechanism for defining abstractions of expressions over parameters. Unlike functions in traditional imperative programming languages, Rosetta functions are pure and side-effect free.

A Rosetta function simply reduces to a value derived from its parameters and items defined in its static scope. Actual parameters that Rosetta functions are applied to cannot be altered by the function application, nor can other symbols in the static scope be altered. Each Rosetta function is simply an encapsulated expression defined over its parameters and items in scope. When evaluated, the encapsulated expression simply reduces to a value that replaces the function application. In this sense, Rosetta functions behave more like mathematical functions than functions and procedures from programming languages.

Throughout this chapter, a simple increment function defined over an integer type is used to demonstrate function properties. Its declaration:

```
inc(x::integer)::integer is x+1;
```

defines a parameter and its type, and range type, and a defining expression. For any expression, a, when `inc(a)` appears in a specification, it can be reduced to a+1.

Direct definition using this style is the simplest and most common mechanism for defining functions, but it is not the only mechanism. Like other Rosetta items, functions have labels, types, and values. While the direct definition mechanism defines all three in one syntactic construct, it is possible to define values and types for functions using more flexible techniques. Function variables, unknown constant functions, anonymous functions, and function values are defined by specifying selected aspects of a function item. This enables the system-level designer to describe function properties without providing a complete implementation when details are not known.

Evaluating Rosetta functions is a two-step process of replacing actual parameters with formal parameters and simplifying the resulting expression. Function evaluation allows replacement of functions by their instantiated values in expressions. Further, evaluation of Rosetta functions is lazy and uses a curried function semantics. As will be discussed later in this chapter, evaluating inc(3) involves replacing inc with its definition and instantiating that definition with 3. Informally:

```
inc(3) == 3+1 == 4
```

6.1 Direct Function Definition

The direct definition approach provides a mechanism for succinctly specifying a signature and optional body for a known function. The format for all direct function definitions is:

f ⟦ [*variables*] ⟧(⟦ *parameters* ⟧) :: T ⟦ **is** e_1 | **constant** ⟧
⟦ **where** e_2 ⟧ ;

Functions defined directly must include a function name, followed by an optional universally quantified parameter list, a parameter list, and a required type. The parameter list defines formal parameters for the function while the type defines its range. The universally quantified parameter list defines universally quantified parameters whose values are determined by inference mechanisms rather than direct instantiation.

To support defining functions at multiple levels of abstraction, Rosetta provides numerous definitional styles. The definition style used in any function declaration is determined by combination of optional **is** and **where** clauses. The **is** clause specifies that the function is constant and an optional expression defines the function's value. If the expression is excluded, the keyword **constant** declares that the function's value is constant although unknown. The **where** clause specifies a property that must be satisfied by any value associated with the function, but does not define a specific function. Both **where** and **is** clauses may be included in the same function.

We say a function is a variable or constant based on whether its definition can vary within a specification. Like all constants, a constant function is a function whose function value does not change. This implies that, evaluated with the same inputs, the resulting value will always be the same. Variable functions are functions whose value may change, implying that, at different times, evaluation with the same parameters may result in different values, depending on the function's current value.

For example, var_fun defines a variable function whose value must be evaluated to a number greater than 5 for all inputs:

```
var_fun(x::integer)::integer where var_fun > 5;
```

Table 6.1 Styles for direct function definition

Definition Style	is Clause	where Clause	Resulting Function Definition
Interpretable	*expression*	*no*	*Function whose type is known, value is known and is constant*
Uninterpretable	**constant**	*no*	*Function whose type is known, value is unknown and is constant*
Qualified Interpretable	*expression*	*yes*	*Function whose type is known, value is a known constant, and satisfies the* **where** *property*
Qualified Uninterpretable	**constant**	*yes*	*Function whose type is known, value is an unknown constant, and satisfies the* **where** *property*
Variable	*no*	*no*	*Function whose type is known, value is unknown and is not constant*
Qualified Variable	*no*	*yes*	*Function whose type is known, value is unknown and is not constant, and satisfies the specified* **where** *property*

There are any number of definitions that satisfy var_fun:

```
const_5(x::integer)::integer is 5;
sqr_plus_5(x::integer)::integer is x*x+5;
```

var_fun is a variable function because either const_5 or sqr_plus_5 could be assigned to it and satisfy its constraints. In contrast, const_5 and sqr_plus_5 are constant functions because their function values are fully specified.

Table 6.1 lists the various definitional styles, showing how to identify them and their implications on the functions they define. The following sections define in detail the definitional styles, when they are used, and what implications they have on specifications.

6.1.1 Interpretable Functions

An *interpretable* Rosetta function is defined by providing a signature followed by an is clause specifying an expression over parameters from the signature and other functions and constants. The signature defines the function name, formal parameters, and a result type. The is clause expression defines how to derive a value for any application of the associated function.

A simple interpretable function is inc for incrementing integers:

```
inc(x::integer)::integer is x+1;
```

In this definition, the function signature, inc(x::integer)::integer, is specified followed by an expression indicated by the **is** keyword, x+1, that defines the calculation performed by the function. The function signature defines a new item that is a function mapping from integer to integer, while the expression defines the actual mapping. Interpreted in the same manner as earlier item definitions, this suggests that inc(x::integer) is an integer whose value is x+1. This intuition is precisely correct — wherever inc(a) appears in a specification, it can be replaced by a+1 with assurance that its evaluation will result in an integer type.

A restriction placed on functions is that the expression following the **is** keyword must only reference items that are declared in its static scope. For example, the following are legal function declarations:

```
squared_sum(x,y::integer)::integer is (x+y)^2;

area(r::real)::real is pi*r^2;

mod_four(i::integer)::integer is
  if i>=3 then 0 else inc(i);

x::integer;
addx(y::integer)::integer is x+y;
```

The squared_sum definition is legal because it only references quantities defined in its parameter list. The area function is legal because it references only quantities defined in its parameter list and the constant pi. The mod_four function is legal because it references only quantities defined in its parameter list and the function definition inc provided earlier. The addx function is legal if x::integer occurs in its static scope. If the declaration of x did not occur in the static scope, the function declaration would be illegal.

Interpretable functions are the most commonly defined Rosetta functions and are so named because they can be interpreted for any parameter values. Specifically, given an interpretable function definition and actual parameters of the appropriate type, the function can always be evaluated with respect to those parameters.

Like any Rosetta definition, inc is an item with an associated type and value. In this case, the type of inc is a function defining a mapping from one integer value to another integer value. The encapsulated expression x+1 along with parameter declarations is the function value associated with the function. Like any other Rosetta declaration, the value of inc must be an element of its type. Because inc is a mapping from integer to integer and the type of the expression x+1 is integer in the context of the parameter declaration x::integer, the function value is of the appropriate type.

Literally, what the inc function definition states is that anywhere in the defining scope inc(a) can be replaced by a+1 for any arbitrary integer values. Whenever any function appears in an expression in a fully instantiated form, it can be replaced by the result of substituting formal parameters with actual parameters and evaluating the resulting expression. Specifically, if the function instantiation inc(3) appears in an expression, it can be replaced by the expression 3+1 and simplified to 4. For this reason, Rosetta functions are frequently viewed simply as encapsulations of expressions.

When a function's value is constant, it is the function value, not the value resulting from applying a function, that is constant: inc is a constant function even though its return value varies. Wherever inc(a) appears in a specification, it will always be replaced by a+1 and never by another expression over a.

An example of an interpretable function definition is the classical definition of factorial:

```
fact(x::natural)::natural is
  if x=0 then 1 else x*fact(x-1) end if;
```

Because all interpretable functions can be evaluated over appropriate typed actual parameters, fact(2), we can provide a naive but useful demonstration of function evaluation. Evaluating fact(2) involves replacing the formal parameter x with the actual parameter 2 in the function expressing and simplifying:

```
1. fact(2)
2.    == if 2=0 then 1 else 2*factorial(2-1) end if
3.    == 2*factorial(1)
4.    == 2*if 1=0 then 1 else 1*factorial(1-1) end if
5.    == 2*1*factorial(0)
6.    == 2*1*if 0=0 then 1 else 2*factorial(0-1) end if
7.    == 2*1*1
8.    == 2
```

Other than using classical definitions for multiplication and subtraction to reduce numerical expressions, the only operations used to perform this calculation are substitution and instantiation.

A semantically precise definition of function evaluation will be provided later. However, this simple heuristic approach of replacing formal parameters with actual parameters and simplifying the result serves as an excellent starting point for understanding function evaluation. ∎

The opcode function pulls the first 4 bits from an input word. Such a function might be used to specify part of a decode operation in a CPU model:

```
opcode(x::word(16))::word(4) is x sub [0..3];
```

opcode is legally defined because it only references its formal parameters.

The circumference function defines the circumference of a circle in the classical manner. One way to define this is to provide a local definition for the value pi within the function definition:

```
circumference(r::real)::real is
  let pi::real be 3.14159 in
    pi*r^2.0
  end let;
```

In this definition, the **let** expression defines a local value for pi that is constant over its scope. This function is legal because it only refers to values defined locally or its parameters.

An alternative definition uses the built-in definition of pi provided in the Rosetta prelude:

```
circumference(r::real)::real is pi^2.0;
```

This definition is also legal because the definition of pi is known to be constant due to the is clause used in its definition.

Finally, the definition of add for complex values might be specified as:

```
add(x,y::complex)::complex is
  (rp(x)+rp(y))+(ip(x)+ip(y))*j;
```

This definition is again legal because it refers to constant values (j) and constant functions and operators (rp, ip, +, and *). ∎

6.1.2 Qualified Interpretable Functions

Adding a **where** clause to an interpretable function adds a constraint that must be satisfied by the function definition. Such functions are *qualified interpretable* functions because they are interpretable, but their definition is qualified by the requirement that they must satisfy their associated **where** clause property.

The definition syntax for qualified interpretable functions is identical to interpretable functions with the addition of a **where** clause following the function definition. The keyword **where** defines a boolean valued expression that must be true for any function definition. This boolean expression is defined over function parameters and constants defined in the scope of the definition. All parameter names are implicitly universally quantified in the **where** expression.

The increment function defined previously can be qualified using a **where** clause as follows:

```
inc(x::integer)::integer is x+1
  where x>=0 implies inc(x)>0;
```

In this case, the **where** clause defines a correctness condition for the function definition specifying that whenever its actual parameter is greater than or equal to zero, interpreting the function results in a value greater than zero. Because all parameters are universally quantified, the **where** clause must be true for any instantiation of the function. The following examples are correct function definitions:

```
squared_sum(x,y::integer)::integer is (x+y)^2
  where squared_sum(x,y)>=0;

area(r::real)::real is pi*r^2
  where r>0 implies area(r)>=0;
```

```
mod_four(i::integer)::integer is
  if i>=3 then 0 else inc(i)
    where mod_four(i)>=0 and mod_four(i)=<3;
```

In each case, the **where** clause can be verified with respect to the interpretable function. If the **where** clause cannot be satisfied by the function definition, then the function is illegal. The following example is an incorrect function definition:

```
inc(x::integer)::integer is x+1
  where inc(x)>0;
```

In this case, the definition of inc cannot be used to verify that all applications are greater than zero. On the contrary, counterexamples are easily found when the argument is negative.

Specifying assertions over a function definition using the **where** clause is important for verification tools that use its definition. Whenever the **inc** definition is used in a specification, the **where** clause can be assumed true. This can be used to assist verification tools by providing facts that need not be derived. Of course, properties proved to be true can be used in this manner. This behavior effectively caches verification results for use in later verification efforts.

The factorial function can be defined with an associated **where** clause that asserts its value must be greater than 0:

```
fact(x::natural)::natural is
  if x=0 then 1 else x*factorial(x-1) end if
    where fact(x)>0;
```

This definition differs from the previous definition because the **where** clause must be satisfied. During type checking and static analysis, the specifier can attempt to verify the **where** clause using type checking, theorem proving, or model checking techniques.

The **where** clause can also be used by functions referencing fact. Given a qualified definition of inc:

```
inc(x::integer)::integer is x+1
  where x>=0 implies inc(x)>0;
```

the **where** clause can be verified. When evaluated with a fact application as its argument:

```
inc(fact(x));
```

the inc function's **where** in combination with the fact function's **where** clause allows verification that the result is always greater than 0. By knowing fact(x)>0, it follows quite simply that fact(x)+1 is also greater than zero because the argument is always greater than 0. Such uses of **where** clauses are exceptionally powerful mechanisms for specifying correctness conditions on functions, terms, facets, and domains. ∎

EXAMPLE 6.3

Qualified Interpretable
Function Definitions

6.1.3 **Uninterpretable Functions**

The **constant** keyword in conjunction with a function declaration provides a mechanism for defining a new function whose specific value is not known, but is constant. Such functions are *uninterpretable* because, although constant, their values are not known and cannot be evaluated in the general case. The signature of the function is defined in the same manner as for other directly defined functions, followed by the **is** keyword, with the **is** expression replaced by the **constant** keyword. For example:

```
addpolyglot(x,y::polyglot)::polyglot is constant;
```

defines a function, addpolyglot, that takes two items of type polyglot and results in a new item of type polyglot without providing a specific definition for the function. The **constant** keyword indicates that the value of addpolyglot is a constant even though its actual value is not known. Other functions can now use the addpolyglot function, knowing that addpolyglot(x,y) will always perform the same function in the context of this definition. Any actual function associated with addpolyglot must obey the definitional rules specified previously.

6.1.4 **Qualified Uninterpretable Functions**

The **where** keyword is used to add constraints to an uninterpretable function definition without defining the function itself. Used with the **constant** keyword, constraints can be defined on a constant function usable in other function and facet definitions. Such declarations are referred to as *qualified uninterpretable* functions. Functions defined in this manner cannot be interpreted in the traditional fashion; however, any instance of the function must obey the constraints. This allows them to be used in analyzing designs. Declaration of qualified uninterpretable functions is identical to uninterpretable functions with the addition of a **where** clause. For example, the following definition makes the addpolyglot function commutative:

```
addpolyglot(x,y::polyglot)::polyglot is constant
  where addpolyglot(x,y) == addpolyglot(y,x);
```

The defined constraint is simply a boolean valued Rosetta expression defined over the parameters of the function. In this case, the constraint states that the order of arguments to addpolyglot is not significant.

EXAMPLE **6.4**

Defining Abstract
Structures
In the definition of an abstract architecture, it is frequently necessary to define operations without defining the specific types operated on or the details of the operation. The following definition provides a mechanism for specifying an operation without significant detail:

```
manipulate_data([T::Type] x::T)::T is constant
  where manipulate_data(manipulate_data(x)) == x;
```

This declaration defines a function, manipulate_data, that must be its own inverse. In this case, we are not specifying the types involved in the operation.

Such functions do not lend themselves to simulation-based analysis, but can be exceptionally useful when defining system requirements before implementation directions are selected. When an implementation for manipulate_data is chosen, design decisions based on the **where** constraints can be enforced on the function, assuring that early decisions are reflected in the implementation. ∎

Using the **constant** and **where** keywords may appear unusual for those accustomed to traditional simulation-style definition languages. The keywords are exceptionally useful when working at high levels of abstraction, where only some of the properties of a function are known. By defining a specific function, too much information may be added to the definition. Using the **constant** and **where** keywords allows specification of desired properties or simply the existence of a function without overspecifying its definition. Details can be added later or discovered during the design process while maintaining the original function definition.

6.1.5 Variable Functions

When a function is defined with no **is** clause, no specific value is associated with the function definition. Functions defined in this manner are called *variable functions* or *function variables*. For example, the following definition defines a function that maps sequences of integer to a single integer:

```
fold(s::sequence(integer)) :: integer;
```

We know that the fold function takes a sequence and generates an integer value, but nothing more. Furthermore, fold is in every sense a traditional variable. Its type is known and fixed while its value may change and cannot be determined by the declaration. The implications of this are that evaluating fold on the same arguments may result in different values as the function's value changes. While the function's value may change, its signature cannot. Specifically, fold will always be a function that maps an integer sequence onto an integer within the scope of this definition.

6.1.6 Qualified Variable Functions

When a function is defined with no **is** clause and a **where** clause, no specific value is associated with the function definition, but any value associated with the definition must satisfy the **where** clause definition. Functions defined in this manner are called *qualified variable functions*. Like variable functions, qualified variable functions are traditional variables. However, at any time the value associated with a qualified variable function must satisfy the boolean expression specified by the **where** clause. For example, the following definition defines a traditional cube root function over real values:

```
cubert(x::real)::real
  where cubert(x)^3 == x;
```

From this definition we know that cubert is a function mapping one real value to another real value. However, without an is clause we do not know what the actual function is. The distinction between this declaration and a variable function declaration is that we know that cubert(x)^3==x for any parameter value. The properties of the qualified variable function are known even when the function value is not. Specifically, we do not know how to calculate the value associated with cubert, but we do know the properties of the evaluated function.

Using a qualified variable function is useful in abstract specification situations, where a function may not be completely defined but properties beyond parameter information and type are needed. The cubert example demonstrates this effectively in that no type condition placed on parameters can provide the same information as provided by the **where** clause.

EXAMPLE **6.5**

Variable Functions

Variable function definitions declare a function without defining its properties. This can be viewed as defining a function signature. Function signatures from the previous examples include:

```
fact(x::natural)::natural;

opcode(x::word(16))::word(4);

add(x,y::complex)::complex;
```

Each signature definition is obtained by simply omitting the is clause from the declaration. Each of these functions is declared, but the specifics of their definition is excluded. Defining function signatures in this manner is an excellent tool for defining abstract system-level properties when only incomplete information is available. Without more information, these variable functions cannot be evaluated. ∎

EXAMPLE **6.6**

Interpretable Functions as
Variable Functions

It is possible to use a qualified variable function to provide the same semantics as provided by interpretable functions, by asserting that evaluating the function **is** equal to the **is** clause expression. Using this technique, the inc function over integer can be defined as:

```
inc(x::integer)::integer where inc(x) = x+1;
```

This definition asserts that any value associated with inc must define a function that, when applied, must result in its argument plus one. The equals sign behaves much like the **is** keyword in that it provides a definition for the value produced by applying inc. The expression x+1 can be used to define a function value for inc. The distinction between this form of a qualified variable function and an interpretable function is that an interpretable function can always be evaluated by virtue of its definition semantics. This form of a qualified variable function may also be evaluated, but that fact arises from the semantics of the **where** clause property. Specifically, the property associates a unique value with each instantiation of inc. Other syntactic forms can achieve the same result. Where interpretable functions may always be evaluated, semantic analysis is required if qualified variable functions can be evaluated. This analysis is highly problematic for automated tools and in most cases cannot be performed. Thus, the interpretable function form is provided to define functions that tools can interpret automatically.   ■

6.2 Function Values and Function Types

Like other Rosetta items, function items have types and values. The direct definition techniques discussed thus far combine the declaration of function items, function types, and function values into a single syntactic construct. However, it is possible to define functions in a manner identical to that for other items by specifying a label, type, and optional value.

6.2.1 Function Values

Function values, also called *anonymous functions*, are the values associated with function items and represent an abstract expression defined over a collection of parameters. Function values correspond to abstractions or lambdas in the lambda calculus and lambda expressions in Lisp dialects, ML, and Haskell. Defined by excluding the function name and encapsulating the definition in the function former, function values provide values for functions without associating them with items. The definition:

```
<* (x::natural)::natural is x+1 *>
```

defines the function value that serves as the value of inc. The parameter list, return type, and expression are identical to the previous inc declaration, but are enclosed in a function former with no function name. The function is identical to inc in every way, but has no associated name. For example, function values and named functions are evaluated in exactly the same manner:

```
1. <* (x::natural)::natural is x+1 *>(1)
2.   == <* ()::natural is 1+1 *>
3.   == 1+1 :: natural
4.   == 2 :: natural
```

In the first evaluation step, x is replaced by 1, resulting in an empty parameter list and an expression with no parameters. The function former delimiters can now be dropped in the second step because the function is nullary. Furthermore, we know that the expression must be the result type of the function. In the third and final evaluation step, the resulting expression is then simplified until it is in normal form, in this case resulting in a natural value.

6.2.2 **Function Types**

Anonymous function signatures can also be defined in a similar manner. The definition:

```
<* (x::natural)::natural *>
```

is the signature of the function defined previously, but does not associate the signature with a name. This definition is called a *function type* because it describes a collection of functions all having the same parameter list and result type. The definition:

```
inc :: <* (x::natural)::natural *>
```

declares a new item inc whose type is a mapping from two natural to natural. This definition is semantically equivalent to the earlier inc signature definition that used the direct definition style. Specifically:

```
inc(x::natural)::natural == inc::<*(x::natural)::natural*>
```

The definition says that inc is of type <*(x::natural)::natural*> or alternatively that inc is a function that maps a natural number to another natural number.

6.2.3 **Alternative Function Item Declaration**

Using the **is** notation, it is possible to define a function using the standard constant definition notation used for other types. By using a function type and a function value in the standard definition syntax, the following definition results:

```
inc :: <* (x::natural)::natural *> is
       <* (x::natural)::natural is x+1*>
```

This definition is equivalent to the preceding direct definition of inc and semantically defines the direct function definition shorthand. Specifically:

```
inc(x::natural)::natural is x+1  =defs
```

```
inc :: <* (x::natural)::natural *> is
       <* (x::natural)::natural is x+1*>
```

The inc item declaration defines it as a function mapping natural to natural. The **is** clause associates a function value with the declared function item in the same manner as for other item declarations. Specifically, the is clause asserts that the inc function is equivalent to the function mapping a natural number to its successor by adding one. Because this declaration uses an **is** clause, inc is a constant function, just as in the earlier definition. Conversely, when the is clause is not present, inc is a variable function.

Although the constant function declaration is equivalent to direct definition, it is highly recommended that direct definition be used for defining functions when the function value does not change. The clumsiness of this definition makes it difficult to read. Furthermore, the direct definition approach is easily recognized as a constant function by language processing tools, making optimization and interpretation simpler.

The formal syntax of the function former used to define anonymous functions and function types is:

<* [[[*variables*]]]([[*parameters*]]) :: *T* [[**is** e_1]] [[**where** e_2]] *>

where *variables* is an optional list of universally quantified parameters, *parameters* is a list of declarations that define formal parameters, *T* is the result type, e_1 is the expression associated with the optional **is** clause, and e_2 is the expression associated with the optional **where** clause. With the exception of not specifying an item name, the form and semantics of the declaration are identical to those of the direct declaration approach. The notation $< * fcn * >$ is referred to as a *function former* because it encapsulates expressions with a collection of local symbols to define a function value.

6.3 Evaluating Functions

Although Rosetta is not an executable language, an operational semantics for function evaluation is defined. Rosetta functions use *curried* function and *call-by-name* semantics, and evaluate *lazily*. This evaluation style is used in lazy languages such as Haskell, Miranda, and Gopher, allowing Rosetta evaluation to treat equality the same way it is treated in mathematics. In Rosetta, = is equality, not assignment. If two terms are equal, then it should be possible to directly substitute one term for another. This is identical to mathematics and quite natural for specifications. The lazy, call-by-name approach facilitates this feature.

The function signatures specified previously all have equivalent forms using function types. The following definitions:

```
fact(x::natural)::natural where fact(x)>0;

opcode(x::word(16))::word(4) is constant;

add(x,y::complex)::complex;
```

are equivalent to:

```
fact::<*(x::natural)::natural*> where fact(x)>0;

opcode::<*(x::word(16))::word(4)*> is constant;

add::<*(x,y::complex)::complex*>;
```

Like other variables, function variables defined in this manner can be constrained by terms in their declaration context. The following facet declares local versions of the previous functions:

```
facet demo(v::input complex)::continuous is
  fact::<*(x::natural)::natural*>;
  opcode::<*(x::word(16))::word(4)*> is constant;
  add::<*(x,y::complex)::complex*>;
begin
  t1: forall(x::integer | fact(x) > 0);
  t2: opcode = <* (x::bitvector(16))::bitvector(4) is
                    x sub [0,..3] *>;
  t3: add = <* (x,y::complex)::complex is
                  (rp(x)+rp(y)+rp(v)) + (ip(x)+ip(y)+ip(v))*j *>
end facet demo;
```

The term t1 places the same assertion on fact as the **where** clause used in its original definition. Specifically, it asserts that every application of fact to a value of type integer is greater than 0. Although semantically equivalent, this syntactic form makes it difficult for tools to recognize that the term asserts the same property. The term t2 defines opcode to be a constant function value equivalent to its earlier defined value. Specifically, the function value specified is the same function used in its interpretable definition. Like the previous term, the semantics of this term makes this definition equivalent to the earlier definition. Similarly, it is difficult for tools to recognize this form without some evaluation or static analysis. Finally, the term t3 defines a new concept called a *quantity*. A non-constant item, namely v, is referenced in its definition. Thus, the actual definition changes over time and is not a constant function or a function value. Such definitions will be explored later and provide an interesting mechanism for specifying simultaneous equations that define a system.   ■

Currying refers to the process of treating any multi-parameter function like a unary function. The Rosetta function f(q,r,s,t) is expanded to its curried form f(q)(r)(s)(t) during evaluation. The curried semantics are quite common and provide a definition that supports partial evaluation and partial instantiation of functions. The technique is particularly powerful when dealing with higher-order functions.

Lazy evaluation implies that expression evaluation occurs only when the expression's value is needed. In traditional imperative languages, parameters to most operators and functions are evaluated before the operator or function. In Rosetta, this is not the case. Terms are evaluated only when their values are needed. An excellent example is the **if** expression, where only the expression associated with the result of evaluating the condition is evaluated.

Call-by-name describes the parameter passing mechanism used to evaluate function application. When passing actual parameters to a function, they are not evaluated prior to resolving the function. The actual terms passed as parameters are substituted directly for the parameters inthe function's associated expression. When the instantiated expression is evaluated, terms that were parameters are evaluated only when their values are needed.

When a complete, interpretable definition exists, repeatedly applying evaluation rules will result in a normal form that is also a value. When an incomplete definition exists, the same process will result in a normal form that is not yet a value. This is one major distinction between a specification language and a programming language. Regardless of whether the result is a value or not, meaningful information results. The normal form is typically simpler and can be used as a partial evaluation result. More importantly, it can be treated as an abstract value that describes a collection of values in abstract interpretation.

6.3.1 Interpretable Functions

Two types of Rosetta functions may always be evaluated — interpretable functions and function values. Anonymous function values defined using an is clause provide an expression that defines the transformation implemented by a function. Thus, an anonymous function value can be evaluated by instantiating its associated expression and simplifying the result. The subset of Rosetta functions defined earlier as interpretable are named because they can always be interpreted. Like anonymous function values, the **is** clause defines an expression that is instantiated and evaluated. Closer examination reveals that the value associated with an interpretable function is an anonymous function value. Thus, the ability to evaluate function values applied to arguments gives us the ability to define interpretable functions.

Unfortunately, it is not possible to determine if an arbitrary function can be evaluated. Many technically uninterpretable functions defined using the **where** clause or using facet terms may in fact be interpretable. Rosetta only guarantees that anonymous function values and interpretable functions may be evaluated. Both definitions are easily recognized by the presence of an **is** clause in their declaration. Without an is clause, it is impossible in to determine if a function can be evaluated.

Evaluating a Rosetta function is a matter of replacing formal parameters with actual parameters in the function-expression and simplifying the resulting expression. The simplest example of function evaluation applies to the add function fully instantiated over literal parameters. A definition of an interpretable add function and an example interpretation result are shown here:

```
add(x,y::integer)::integer is x+y;
```

```
1. add(1,2)
2.    == 1+2
3.    == 3
```

In the evaluation, formal parameters are replaced by actual parameters in the add definition and the resulting expression is evaluated.

Closer examination of the instantiation and evaluation process reveals a powerful capability for currying and lazy evaluation that is essential in system-level design. Let us first re-examine the evaluation of the add function just performed in simplified fashion, looking more carefully at the evaluation process:

```
1. add(1,2)
2.    == <* (x,y::integer)::integer is x+y *>(1,2)
3.    == <* (y::integer)::integer is 1+y *>(2)
4.    == <* ()::integer is 1+2 *>
5.    == 1+2 :: integer
6.    == 3 :: integer
```

When evaluating a function, the function item is replaced by its value. In the previous section we showed that every interpretable function item has an associated function value. Here, in step 2, the add item is replaced by its function value whose definition maps two integer parameters onto their sum. In step 3, the formal parameter, x, is replaced by the actual parameter, 1. The formal parameter is eliminated from the signature in this process and the result is a new function of one integer parameter applied to the remaining actual parameter. In step 4, the formal parameter, x, is replaced by the actual parameter, 2, again eliminating the formal parameter.

The function resulting from step 4 is nullary — having no remaining parameters. Its associated expression is fully instantiated and can be removed from the function value former. The empty parameter list is dropped and the function result type is associated with the resulting expression. Generalizing, we have the following rule for any nullary function:

$$<* ()::T \text{ is } e *> \stackrel{\text{defs}}{=\!=} e :: T$$

where T is a type and e is an expression. Note that the definition can be used inversely to place an expression in a nullary function. This rule can be further generalized to define an abstraction rule. Given a variable v not free in e, the following rule can be defined:

$$<* (v::T_p)::T \text{ is } e *> \stackrel{\text{defs}}{=\!=} e :: T$$

The right side of the definition is the same, with the left side adding a parameter that is not free in e. What this means is that v does not appear in e unless it is declared somewhere in e. If that is the case, the value of v can have no effect on the evaluation of e and can be dropped during evaluation or added during abstraction.

The previous method for evaluating a multi-parameter function by successively applying the function to each parameter defines a curried function evaluation process, a classical evaluation semantics. It is at the same time simple and exceptionally powerful.

It is interesting to observe that the resulting process produces identical results if the parameters are instantiated in reverse order:

```
1. add(1,2)
2.    == <* (x,y::integer)::integer is x+y *>(1,2)
```

```
3.   == <* (x::integer)::integer is x+2 *>(1)
4.   == <* ()::integer is 1+2 *>
5.   == 1+2 :: integer
6.   == 3 :: integer
```

or at the same time:

```
1. add(1,2)
2.   == <* (x,y::integer)::integer is x+y *>(1,2)
3.   == <* ()::integer is 1+2 *>
4.   == 1+2 :: integer
5.   == 3 :: integer
```

This process, called *beta-reduction,* allows great flexibility in dealing with function evaluation. Not only is the instantiation process simple, it supports currying without enhancement of the function or evaluation definitions. Simply substitute and simplify — a process widely used in mathematics. Note that in subsequent evaluation examples, the nullary function and result type will frequently be dropped for brevity.

Nested functions work identically by applying the replace and simplify approach:

```
inc(add(4,5))
```

Rather than evaluating in the traditional style, expand the function definitions first. Using the definitions of inc and add results in the following anonymous function evaluation process:

```
1. <*(z::integer)::integer is z+1*>
      (<*(x,y::natural)::natural is x+y*>(4,5))
2.   == <* ()::integer is <* (x,y::natural)::natural is x+y *>(4,5) +1 *>
3.   == <* (x,y::natural)::natural is x+y *>(4,5) + 1 :: integer
4.   == <* (y::natural)::natural is 4+y *>(5) + 1 :: integer
5.   == <* ()::natural is 4+5 *> + 1
6.   == 4+5 :: natural + 1 :: integer
7.   == 10 :: integer
```

In step 1, the function instantiations are replaced by their definitions. In step 2, the formal parameter, to inc is replaced by the actual parameter, which in this case happens to be an instantiated function. However, the process is unchanged — replace formal parameters with actual parameters and eliminate them from the signature in each step. In step 3, the function former associated with inc is eliminated, as there are no parameters to that function. Now evaluation turns to the add function, where the process is identical to that performed in the previous example.

The same result occurs regardless of the order of replacement and evaluation. The following process shows a different order resulting in the same result:

```
1. <* (z::integer)::integer is z+1 *>(add(4,5))
2.   == <* ()::integer is add(4)(5)+1 *>
```

```
3.    == <*(x,y::natural)::natural is x+y *>(4)(5)+1 :: integer
4.    == <* (y::natural)::natural is 4+y *>(5) + 1 :: integer
5.    == <* ()::natural is 4+5 *> + 1 :: integer
6.    == 4+5 :: natural + 1 :: integer
7.    == 10 :: integer
```

Function evaluation becomes a simple matter of replacement of actual parameters by formal parameters and simplifying. Because this process can be performed in any order and can be stopped at any step, a wide range of function evaluation possibilities become available. Thus far, examples of function evaluation have included only fully instantiated, constant functions. The definition of function evaluation does not preclude designers or tools to evaluate functions partially, either by leaving parameters uninstantiated or by simply halting a simplification process before it completes. As shall be seen, partial evaluation provides for definition and evaluation techniques that directly support high-level specification and analysis.

6.3.2 Curried Function Evaluation

As noted previously, Rosetta uses a curried function semantics to define evaluation. Thus, every Rosetta function can be expressed as a single argument function. This may seem odd, but it results in an exceptionally simple yet expressive mechanism for defining function evaluation. The addc function:

```
addc(x,y::integer)::integer is x+y;
```

can be expressed equivalently as:

```
addc(x::integer)::<*(y::integer)::integer*> is
    <* (y::integer)::integer is x+y *>;
```

Looking carefully at this definition, what was a function of two arguments is now a function of one argument that results in a function of one argument. Specifically, the new function is a unary function over items of type integer that results in another unary function that maps items of type integer to type integer. But how is this equivalent to a two-argument function?

To understand the equivalence, examine the evaluation of addc(5)(4), the equivalent of evaluating add(5,4) using our earlier definition. We start by evaluating addc(5), remembering that Rosetta function evaluation is simply substitution of formal parameters for actual parameters:

```
1. addc(5)
2.   == <* (x::integer)::<*(y::integer)::integer*> is
            <* (y::integer)::integer is x+y *> *>(5)
3.   == <* <* (y::integer)::integer is 5+y *> *>
4.   == <* (y::integer)::integer is 5+y *>
```

The evaluation progresses by first replacing the addc function with its value in step 2. Following this is the actual evaluation, replacing the formal parameter x with the actual parameter 5. In step 4, the outer function former is dropped because no parameters remain. The result is a new function value that adds 5 to its argument. Because this form cannot be evaluated further, it is called a *normal form*. In this case the normal form is also a *value*, specifically a function value. Not all normal forms will be values.

Now we can evaluate addc(5)(4) by applying the result of the previous evaluation of addc(5) to the formal parameter 4:

```
1. addc(5)(4)
2.   == <* (y::integer)::integer is 5+y *>(4)
3.   == <* ()::  integer 5+4 *>
4.   == 9 :: integer
```

In step 2, addc(5) is replaced with the result obtained in the previous evaluation and applied to 4. In step 3, y is replaced with 4, and finally the result is evaluated and 9 results, as anticipated.

It is not necessary to evaluate the results of calling the single-argument add function. Specifically, it is perfectly legitimate to call the function and treat the result itself as a function. The evaluation of addc(1) has the following form:

```
1. addc(1)
2.   == <*(x::real)::<*(y::real)::real is x+y*> * >(1)
3.   == <*(y::real):real is 1+y*>
```

This is exactly the definition of the function value previously associated with inc; inc can be now defined using this evaluation of addc:

```
inc::<*(x::integer)::integer*> is addc(1);
```

addc(1) evaluates to a function value that is the same as the definition of an increment function. Thus, it is perfectly legal to use the result of applying add as the value associated with the inc function definition.

Having discussed the evaluation of addc, it is important to remember that, semantically, addc and add are the same function. The only difference is that addc must be treated as a single parameter function while add can be treated as a two-parameter function or as a curried function. For this reason, using the add approach of specifying multiple parameters in the signature is more flexible and highly encouraged. We can always substitute and simplify as in previous examples, but the semantics of this process is defined by the curried approach discussed here. The definition for inc can use add instead of addc with virtually no modification:

```
inc::<*(x::integer)::integer*> is add(1);
```

The inc function is thus defined as a mapping from integer to integer whose value is the result of curried evaluation of the add function over the value 1.

6.3.3 Uninterpretable Functions

When thinking about evaluating uninterpretable functions of any kind, little can be asserted. Evaluation cannot be guaranteed by syntactic analysis. Furthermore,

EXAMPLE 6.8

Partial Evaluation

no algorithm exists to determine whether a function can be evaluated for general functions. As a rule, if evaluation is necessary, use an interpretable function definition or associate the function with a constant in its declaration.

Partial evaluation is the process of taking a function and instantiating only a subset of its parameters. Curried functions provide the primitive concepts behind partial evaluation. Using Rosetta's curried function semantics, it is possible to define a form of partial evaluation where any variable may be instantiated.

Consider the following definition of f over real numbers:

```
f(x,y,z::real)::real is (x+y)/z;
```

Like any function, f is applied by replacing formal parameters with actual parameters in the definition and evaluating the result. Specifically, f(1,2,3) is evaluated as follows:

```
1. f(1,2,3)
2.   == <*(x,y,z::real)::real is (x+y)/z*>(1,2,3)
3.   == <*(1+2)/3*>
4.   == 1
```

We can make f an average function if the z parameter is instantiated with a constant 2. This is accomplished using the following notation:

```
<* (x,y::real)::real is f(x,y,2) *>
```

The definition of average follows directly as:

```
average::<*(x,y::real)::real is
    <*(x,y::real)::real is f(x,y,2)*>
```

Specifically, average is defined as a function from two real values to a third real value. The function value associated with average is the function f with its z parameter instantiated with 2. Partial evaluation is an exceptionally useful tool for defining and evaluating functions. The preceding example shows a simple case where a new function definition is produced through partial evaluation. The same technique can be used to make analysis simpler and faster as well as to perform analysis of new definitions. ∎

Knowing this, function evaluation still occurs in a manner identical to that for interpretable functions. The following facet definition declares and defines a function, inc, in a highly general manner:

```
facet counter(x::output integer)::state_based is
  inc::<*(x::integer)::integer*>
begin
  t1: forall(x::integer | inc(x) == x+1);
  t2: s'==inc(x);
end facet counter;
```

This facet declaration defines a variable function and uses a new concept (defined in Chapter 7) called a *quantifier* in term t1 to provide a definition. The quantifier asserts that for any value, x::integer, inc(x) is equal to x+1. This should be immediately recognized as equivalent to previous definitions of the increment function. In term t2, inc(s) defines the value of the next state, s'.

Semantically, the value of inc(s) is known for any value of s. However, expecting a tool to determine this and evaluate the function appropriately is not appropriate. When defining systems requirements, such general specifications can play an important role when working at high levels of abstraction, as shall be seen in subsequent discussions.

6.3.4 Qualified Functions

Addition of a **where** clause makes an assertion on function evaluation. Specifically, the **where** clause specifies a boolean condition the function must always satisfy. If the function does satisfy the **where** clause, then evaluation proceeds as if the clause did not exist. If the function does not satisfy the clause, then evaluation results in **bottom**. The **where** clause can be associated with both interpretable and uninterpretable functions and plays an important role in evaluation and analysis.

Qualified interpretable functions are a form of interpretable function and thus can be evaluated in the same manner, with a change to reflect the presence of the **where** clause. Consider a definition of inc that includes a **where** clause:

```
inc(x::integer)::integer is x+1
    where inc(x) > 0;
```

The **where** clause defines a condition on inc that asserts the result of incrementing is greater than 0. If we instantiate the **where** clause in the same manner as the function, the resulting evaluation informally looks as follows:

```
1. inc(2) where inc(2) > 0
2.   == <* (x::integer)::integer is x+1 *>(2)
            where <* (x::integer)::integer is x+1 *>(2) > 0
3.   == <* ()::integer is 2+1 *> where <* ()::integer is 2+1 *> > 0
4.   == 2+1 :: integer where 2+1 :: integer > 0
5.   == 3 :: integer where 3 :: integer > 0
6.   == 3 :: integer where true
7.   == 3 :: integer
```

In this case, the **where** clause is clearly true and the result of the function evaluation is the same as a simple interpretable function. If we evaluate the same function on a different value, inclusion of the **where** clause plays a more substantial role:

```
1. inc(-2) where inc(-2) > 0
2.   == <* (x::integer)::integer is x+1 *>(-2)
            where <* (x::integer)::integer is x+1 *>(-2) > 0
3.   == <* ()::integer is -2+1 *> where <* ()::integer is -2+1 *> > 0
4.   == -2+1::integer where -2+1::integer > 0
5.   == -1::integer where -1::integer > 0
6.   == -1::integer where false
5.   == bottom
```

In this case, the **where** clause is false and the result of the function evaluation is **bottom**.

The **where** clause plays a more important role in the definition of uninterpretable or variable functions by providing information about its eventual definition. The Rosetta prelude defines the square root function as follows:

```
sqrt(x::complex)::complex where sqrt(x)^2 == x;
```

The function does not specify a mechanism for calculating the square root, but asserts a condition on any implementation of it. Static analysis tools can use this information to perform various types of analysis. More importantly, the **where** clause provides implementation direction to a tool user or a specifier who might require a specific function instance. If a function value is associated with sqrt, the clause defines its requirements.

The **where** clause can be treated as a correctness condition on the function, or as an assertion for use by tools interpreting the function. Treated as a correctness condition, tools may either formally verify that an interpretable function satisfies its requirements or, as a program assertion, that it is evaluated when the function is evaluated. Type checkers will attempt to statically verify the **where** clause in an attempt to predict and avoid evaluation errors. When static verification cannot be performed, evaluation-time checking must be instituted.

Because function application is undefined whenever its associated **where** clause is violated, static analysis tools, including type checking systems, may assume that it is satisfied when analyzing surrounding specifications. This has many benefits, including assertion of function properties in a manner similar to that for type assertions. When a property is assumed or verified, the **where** clause can be added to assist verification tools by asserting properties.

6.4 Universally Quantified Parameters

Universally quantified parameters in function definitions provide Rosetta functions with a form of polymorphism by allowing type inference. When function applications are evaluated or statically checked, parameter values are specified by instantiating them with expressions. In contrast, universally quantified parameter values can be inferred rather than explicitly specified.

EXAMPLE 6.9

Qualified, Uninterpretable Functions

A principle use for qualified function definition is defining requirements for functions whose implementations need not be specified. Such functions are frequently defined in the prelude, where the definitions of basic Rosetta functions are defined without regard to specific implementations. One such function is the choose operation over sets. Recall that choose(s) can evaluate to any element from s. If we use interpretable functions to define choose, we are required to specify a selection mechanism. Using the **where** clause, it is easy to define requirements without specifying mechanism:

```
choose[T::type](s::set(T))::T is constant
  where choose(s) in s;
```

This definition of choose provides a precise definition without specifying how the operation is implemented. ■

A polymorphic form of the inc function provides a motivating example. The definition used thus far has the following form:

```
inc(x::integer)::integer is x+1;
```

We also know that inc is defined for subtypes of integer and for some super-types such as real. Subtypes present little problem, as we can simply use the traditional definition. However, supertypes cannot be handled by this definition. Furthermore, when a subtype of integer is used, the result will always be of type integer as specified by the function signature.

One solution to this is to make the operand type a parameter to inc:

```
inc(T::type; x::T)::T is x+1
  where T =< number;
```

This new definition defines inc for any type that is a number, allowing the specifier to indicate the specific type. We now have a polymorphic inc that can be used in the following manner:

```
inc(natural,3) == 4::natural
inc(real,3.0) == 4.0::real
```

This new form requires the type to be explicitly specified in the function application. Although this solves the original problem, it is still somewhat clumsy. Imagine requiring the type contained in a sequence to be specified each time an operation is applied.

The solution is provided by universally quantified parameters whose types are inferred by evaluation and static analysis tools. In the following definition, T is a universally quantified variable that represents a type:

```
inc[T::type](x::T)::T is x+1
  where T =< number;
```

When this new inc is applied, the value of T is inferred rather than explicitly specified. For example, in the application inc(3), T will be inferred to have the value posint implying:

```
inc(3)::posint == 4::posint
```

Based on the value 3, the Rosetta type system constrains T to be the most specific type possible — in this case, posint. Similarly, in the application inc(3.4), T is constrained to posreal as the most specific type:

```
inc(3.4)::posreal == 4.4::posreal
```

The type system can be forced down a specific constraint path by either directly or indirectly constraining T. Directly constraining T simply involves providing an actual parameter in the traditional manner:

```
inc[real](3.4)::real == 4.4::real
```

Alternatively, type ascription can be used to specify the type of the actual parameter, indirectly constraining T:

```
inc(3.4::real)::real == 4.4::real
```

An interesting problem occurs when T is constrained to a type that is not closed under addition, such as negreal. A problematic example would be:

```
inc(-0.5)::negreal == 0.5::negreal
```

This case is prevented because the type signature of the addition operator prevents T from being constrained to negreal due to the presence of 1 in the addition expression. T can only be constrained to real. If the application attempts to constrain the type indirectly or directly to negreal, the result is a type error.

In contrast, consider a definition of inc using **top** as its domain and range:

```
inc_top(x::top)::top is x+1;
```

Like the polymorphic inc using universally quantified parameters, inc_top can be applied to any value. However, the similarities stop there. Because "+" is defined for any number type, if the type value associated with T satisfied inc's original **where** clause, the addition could be performed. Additionally, nothing can be said about inc_top's result type. If it is used in the context of other numbers, it will not satisfy typing rules. The lesson here is to use universally quantified parameters for polymorphic functions. Using **top** results in functions that are far more difficult to use. This lesson will hold wherever we see a need for polymorphism.

All of the examples thus far deal with using universally quantified parameters with types to implement polymorphism. They can also be used for other purposes, such as specifying operations over variable-length bitvectors or using type inference to perform a kind of meta-specification. The key is that the Rosetta evaluation system must be able to infer a value for universally quantified parameters statically. Evaluation-time information cannot be used to resolve the parameter's value.

We will see universally quantified parameters appear in the signatures of facets, components, and other Rosetta structures. In every case they play the same role as they do in function definition, providing a controlled form of parametric polymorphism. The same techniques will be used to determine their values either by specific instantiation or from implied constraints.

Higher-Order Functions

A *higher-order function* **is simply a function** that accepts other functions as arguments. As presented in Chapter 6, Rosetta functions are first-class items having labels, types, and values like any other declared item. Thus, it is possible to define a higher-order function simply by including a parameter in with a function type in a function's signature. Consider the simple example of a selection function for choosing between demodulation schemes in a receiver:

```
demod(dem1,dem2::<*(s::real)::real*>;
    modType::boolean; x::real)::real is
  if modType then dem1(x) else dem2(x) end if;
```

In this function, x is the input signal that will be demodulated and modType indicates what modulation type to select. The parameters dem1 and dem2 are functions that map real values to real values and represent the two alternative demodulation schemes. When evaluated, the function uses its boolean input to select from the two demodulation schemes. The advantage to this approach is that demodulation types need not be known when the function is developed. The designer may select a general architecture for the design while deferring specifics until a later point in the design process process. The demodulation function can be instantiated as follows:

```
demod(qpsk,vsb,true,v)
```

where v is the input signal and qpsk and vsb are functions that perform demodulation.

The approach becomes more powerful when combined with curried function evaluation. Using curried function evaluation, it is necessary to provide values for only a subset of a function parameters. The following definition shows an example of defining a specific demodulation paradigm using the functions defined above:

```
demod1::<*(modType::boolean;x::real)::real*> is
  demod(qpsk,vsb);
```

When evaluated, demod(qpsk,vsb) results in a new function that evaluates an input signal with either qpsk(x) or vsb(x), depending on the boolean select value, modType. This resulting function is used as the value for function demod1. The demod function provides an architecture for demodulation functions that can be instantiated for specific situations.

We can take the example one step farther and instantiate demod1 to be a qpsk demodulator exclusively:

```
qpsk1::<*(x::real)::real*> is
  demod1(true);
```

Most applications of higher-order functions are transparent to the Rosetta user. Specification functions such as quantifiers, set operations, and map and filter operations represent applications of higher-order functions that are used without fanfare in the base language. However, the higher-order technique is available to users and represents a powerful specification technique.

7.1 Domain, Range, and Return Functions

The core of Rosetta's built-in higher-order functions includes functions for extracting the domain, range, and return type from a function definition. Described in Table 7.1, dom, ran, and ret provide capabilities for extracting domain, range, and return type from a function, respectively. The dom function accepts a function as its only argument and returns the domain of that function. For unary functions such as inc, this is simply a process of returning the type associated with the input parameter. If inc is defined in the traditional fashion:

```
inc(x::natural)::natural is x+1;
```

the domain of inc is found by using the dom function:

```
dom(inc) == natural;
```

ret is the return type of the function. Using inc again as an example, the return type is found by evaluating ret on the function:

```
ret(inc) == natural;
```

Table 7.1 Higher-order functions for accessing properties of function definitions

Operation	Format	Meaning
Domain	dom(F)	*Domain of a function*
Range	ran(F)	*Range of a function*
Return Type	ret(F)	*Return type of a function*

Like the domain function, the return type function is realized by simply examining the function definition.

The ran function accepts a single, arbitrary function as its only argument and returns the actual range of the function. Unlike the return type function, the range function evaluates to the function's range by assembling the results of applying the function to each element of its domain type. Determining the range of inc is achieved by evaluating inc on every value of its domain. Thus, the range of inc is the natural numbers without zero. Formally:

    ran(inc) == {x::dom(inc) | inc(x)} == posint;

where posint is the name of the type containing all natural values except 0. It is known that for any function, the range must be contained in the return type:

    forall(f::function | ran(f) =< ret(f));

The domain and range functions present a greater challenge when dealing with functions of arity other than 1. It is possible to define a function with no arguments, often called a nullary function. During function evaluation, nullary functions are always reduced to expressions by removing the function former. Specifically:

    <* ()::*T* **is** *e* *> == *e*

This is perfectly legal, as the purpose of the function former brackets is to define the scope of function parameters. With no parameters, there is no need to define scope. The domain of all nullary functions is the empty type containing no values. The range of a nullary function is the result of evaluating its associated expression. Thus:

    dom(()::natural **is** <* 3+2 *>) == {}
    ran(()::natural **is** <* 3+2 *>) == {5}

Using these identities, one can define evaluation of a fully instantiated function as taking its range after instantiating and removing parameters. If all arguments to a function are known, then the range of that instantiated function is the same as evaluating the function.

EXAMPLE **7.1**

The Workhorse of
Higher-Order Functions

Although seemingly quite innocent, the range function is the workhorse of the built-in higher-order functions. Virtually all the higher-order functions we will see involve calculation of range in their definition. Although it is not necessary to understand such definitions to use the functions, some understanding can provide valuable insight into the definition of other higher-order functions.

In reality, ran behaves like an image function. Recall that an image function takes a collection of items such as an array or list and applies a function to each element. The difference between this mapping function and the Rosetta range function is that the initial set of values to be mapped is the domain of the specified function, while the function mapped to each element is the function definition.

Given that we would like to apply the function f(x::D)::R to the collection of items d, we can use the image function as follows:

```
image(f,d);
```

The ran function is the same, except it takes each element of a function's domain and applies the function to it. Using image again and dom to obtain the range of f, this function can be implemented as:

```
image(f,dom(f))
```

This application of image is semantically equivalent to:

```
ran(f)
```

Note that when actually defining Rosetta semantics, ran is the primitive function and image is defined from it.   ∎

The domain of functions with more than one parameter is defined as the type of the first parameter. This follows directly from the curried semantics, where every function can be reduced to a unary function. The range of functions with more than one parameter is defined as the type of the function resulting from curried evaluation with respect to the first parameter. Finally, the return type is the type syntactically specified as the return type of the function.

Given a find function that determines if a value is in a sequence of values:

```
find(x::natural,y::sequence(natural))::boolean is
  if y/=null then
    if x=y(0) then true
       else find(x,(y sub [1,..#y-1]))
    end if
    else false
end if;
```

domain, range, and return type are defined as:

```
dom(find) == natural
ran(find) == <*(y::sequence(natural))::boolean*>
ret(find) == boolean
```

The domain is the type of the first parameter and the return type is the type associated with the function. If this function were partially evaluated specifying only the first parameter, the resulting function type would be a mapping from an array of naturals to a boolean value. Thus, the range of the find function is the function type specified in the previous definition.

EXAMPLE 7.2

Defining image Using ran

The primary reason for including the ran function in the Rosetta definition is its usefulness in defining other functions. One such function is image, where a function is applied to elements of a set. In this definition, we will first construct a a function from image(f,S) and evaluate ran on the result. What we would like to apply ran to is a function that accepts the elements of S as its inputs and evaluates to the same value as f. Such a function can be defined as:

```
<*(x::S)::ret(f) is f(x)*>
```

This function's domain is the set S used as one parameter to image. Its value is f evaluated on its input. To define image we simply call ran on the result:

```
image(f,S) == ran(<*(x::S)::ret(f) is f(x)*>)
```                                                                    ∎

7.2 Alternate Higher-Order Function Notation

The functions dom, ran, and ret are defined to return elements of a function definition. Thus, it is natural to call these functions using functions as parameters. When thinking about image and mapping functions, Iminimum, maximum, and quantifier functions, the common practice is to start with a set of items and perform a specified operation on that set. Mapping functions apply a transformation to each element of a set, minimum and maximum return the appropriate value from a set, and quantifiers check to see if a property holds for all or some values from a set. Although each of these properties can be expressed using a higher-order function, it can become difficult to read and understand.

Rosetta introduces a standard notation to allow a more natural expression of higher-order properties involved in quantification associated with sets. The following notation is equivalent to finding the minimum value generated by inc in the earlier example:

```
min(x::natural | x+1)
```

Using this notation, the domain and value expression of the function are separated in a manner consistent with traditional notations. The interpretation of this notation is to take all elements of natural, apply the expression x+1, and return the minimum resulting value. The mapping from this notation back to the higher-order function is a simple syntactic manipulation. If S is a **set** and e is an expression defined over elements of S, then the following relationship is always true:

$$\texttt{min(x::}S \texttt{ | } e) \stackrel{\text{defs}}{=} \texttt{min(<*(x::}S\texttt{)::}T \texttt{ is } e\texttt{*>)}$$

Here the elements of the set are pulled out and used as a type in the function former. The notation:

```
min(x :: natural | x)
```

specifies the minimum value in the set of natural numbers.

What makes this notation powerful is that natural can be replaced by any set and Rosetta forms the higher-order function. If the collection can be identified and the expression defined, the preceding notation can be used to accomplish

Table 7.2 Second-order quantifier functions

Operation	Format	Meaning
Minimum and Maximum	`min(F),max(F)`	*Minimum or maximum value from a function's range*
Comprehension	`sel(P)` `sel(x::T|P(x))`	*Elements from a predicate's domain that satisfy its definition*

Universal and Existential Quantifiers	`forall(P),`
	`forall(x::T|P(x)),`
	`exists(P),`
	`exists(x::T|P(x))`
		True if a predicate is true for all or one of its domain values, respectively

the higher-order specification task. If additional single-parameter higher-order functions are defined, the same notation applies. Semantically, the notation adds nothing, but it significantly enhances the readability of specifications.

7.3 Minimum and Maximum

The primary functions used to define quantifiers are the minimum and maximum functions (Table 7.3). Each of these functions takes as its single argument an arbitrary function, finds the range of that function, and returns the minimum or maximum value in the range, respectively. In other words, the minimum function returns the smallest value a function can produce, while the maximum produces the largest.

The **min** function takes a function and evaluates that function on every possible domain value and selects the minimum value. This can be viewed as taking the function's range and returning the minimum of the range values. A simple function example is the identity function:

```
id(x::natural)::natural is x;
```

This function is equal to its single natural number argument. Applying the **min** function returns its smallest possible value, or, in this case, the smallest possible natural number:

```
1. min(id)
2.    == min(<*(x::natural)::natural is x*>)
3.    == 0
```

The domain of the argument function is the type natural. The range of the argument function is the expression applied to each element of the

Table 7.3 Applying built-in higher-order functions to the contents of sequences

Operation	Syntax	Meaning
Maximum and Minimum	**max**(x::~S | F(x)) **min**(x::~S | F(x))	*Maximum or minimum value* *from the set resulting in F* *applied to a sequence*
Range	ran(x::~S | F(x))	*The set containing F applied* *to all values in a sequence*
Comprehension	**sel**(x::~S | P(x))	*Filtering contents of a* *sequence*
Universal and Existential quantification	**forall**(x::~S | P(x)) **exists**(x::~S | P(x))	*Universal and existential* *quantification over contents* *of sequences*

domain — specifically, the natural numbers. The **min** function then returns the minimum value associated with natural or 0.

Similarly, the **min** function can be applied to inc where the expression associated with the function is x+1. Specifically:

```
1.  min(inc)
2.     == min(<*(x::natural)::natural is x+1*>)
3.     == 1
```

The **max** function is defined similarly and operates in the same manner, except it returns the maximum value associated with a function's domain, rather than the minimum value.

7.4 Quantifiers and Comprehension

The concept of a quantifier from traditional logic allows specifiers to make statements about sets of items. The universal quantifier, called **forall**, allows the specifier to claim that a property is true for every element of a collection. Similarly, the existential quantifier, called **exists**, allows the specifier to claim that a property is true for at least one element of a collection. Although mysterious in many languages, Rosetta quantifiers are simply higher-order functions defined over predicates. In essence, we will take the range of a collection and use that range to determine if a quantifier holds. If the range of a predicate does not contain **false**, the predicate is true *for all* elements of its range. Similarly, if the range of a predicate contains the **true** value, *there exists* a value from the range for which the predicate holds.

Unfortunately, quantifiers and comprehension operators can be somewhat mysterious in their application and definition. In Rosetta, these operators are simply higher-order functions whose definitions do not differ substantially from simple minimum and maximum functions. Quantifier and comprehension functions and the other predefined higher order functions have the same form. The former can be instantiated with functions as their single argument:

```
forall(<*(x::S)::T is P(x)*>)
exists(<*(x::S)::T is P(x)*>)
sel(<*(x::S)::T is P(x)*>)
```

or using the equivalent, more readable mathematical notation:

```
forall(x::S | P(x))
exists(x::S | P(x))
sel(x::S | P(x))
```

where S is a set or type, T is a type, and $P(x)$ is a boolean expression defined over x. The **forall** and **exists** functions determine if P holds for all or some of the values in S, respectively. The **sel** function selects elements of S for which P holds.

Consider the following application of **forall** to determine if a particular set contains only values greater than zero:

```
forall(x::{1,2,3,5} | x>0)
```

Applying the expression x>0 to each element of the input collection, the resulting set becomes:

```
{true,true,true,true}
```

Thus, the expression is **true** for each element of the input collection, meaning that the result of applying **forall** is also true. Assuming the input set is {1,2,3,0} demonstrates the opposite effect. Here, the result of applying the expression to each element of the collection is:

```
{true,true,true,false}
```

Because one value is **false**, the **forall** expression evaluates to **false**.

Rosetta interprets the original definition by forming a function argument to **forall**. In this case, the function has the following form:

```
forall(<*(x::{1,2,3,5})::boolean is x>0 *>)
```

Here, the domain of the argument function is the set {1,2,3,5} and the result expression x>0. To determine the range of the argument function, x>0 is applied to each element of domain collection giving the result presented previously. The **exists** function works identically, but is **true** when at least one of the evaluation results is **true**, rather than all. For example:

```
exists(x::{1,2,3,-1} | x>0)
```

is **true** because the result of applying the expression is {**true,true,true,false**} and one element of the input collection is greater than 0. In contrast:

```
exists(x::{1,2,3,0} | x<0)
```

produces {**false,false,false,false**}. None of the input collection is less than 0 and the **exists** function evaluates to false.

EXAMPLE 7.3
Using **forall** and **exists** as **and** and **or**

One way to think of **forall** and **exists** is as general purpose **and** and **or** functions. Given some set of number values S and a predicate p that determines if a value is greater than 5, the following definitions hold:

```
p(x::integer)::boolean is p>5;

forall(x::image(p,S) | x)
exists(x::image(p,S) | x)
```

Because p is a **boolean** function, the set that x is selected from is **boolean**. Thus, x can simply be checked without transformation. The **forall** function fails if it encounters a single **false** value, while **exists** succeeds if it encounters a **true** value.

These applications are semantically equivalent to:

```
forall(x::S | p(x))
exists(x::S | p(x))
```

The **sel** function provides a mechanism for selecting from a set values that satisfy a property, and returns them as a set. In traditional logic, the selection function is referred to as comprehension and is used as a primary method along with extension for constructing sets. Consider the following example, where **sel** is used to filter out all elements of a set that are not greater than 0:

```
sel(x::{1,2,3,0} | x>0)
```

In this case, the result of evaluating the **sel** function is the set {1,2,3}, or the subset of the input set that is greater than zero. Like the previous higher-order functions, this form of the **sel** function creates a function from the input arguments and operates on the range of that function. For this example, the equivalent form using a function argument is:

```
sel(<* (x::{1,2,3,0})::boolean is x>0 *>)
```

EXAMPLE 7.4
Defining **forall** and **exists**

It is interesting to note here that **forall** and **exists** behave identically to **min** and **max** for **boolean** valued functions. The **min** and **max** functions applied in the same way would result in the same outcome with the following axiom defined:

```
true > false
```

With this axiom:

```
exists(x::1,2,3,-1 | x>0) == max(x::1,2,3,-1 | x>0)
forall(x::1,2,3,-1 | x>0) == min(x::1,2,3,-1 | x>0)
```

In both cases, the result of applying x>0 to each element is {**true,true,true, false**}. Interpreting these values under the assumption that **true>false**, **true** is the maximum value in the result and **false** is the minimum. These values correspond to the desired result of applying **exists** and **forall**, respectively. Thus, **forall** and

exists simply provide more meaningful names for **min** and **max**. Furthermore, **max** and **min** are referred to as quantifiers in addition to the traditional **forall** and **exists** operators.   ■

EXAMPLE **7.5**

Using the **sel** Function

Just as ran can be used to define image, **sel;** can be used to define **filter**:

    filter(p,S) = **sel**(x::S | p(x))

This application of **sel** takes every element of S, applies p, and keeps the value from S if p is satisfied.   ■

EXAMPLE **7.6**

Using **sel** to Define Types

Possibly the most common use of the **sel** function is to define new types. Chapter 8 discusses how any set can be used as a Rosetta type. All types defined in this and previous chapters are in fact sets of items. The **sel** function is used extensively to define subtypes of existing types. For example, natural is the subtype of integers that includes zero and the positive values. The definition of natural type is achieved using the following definition:

    natural::**subtype**(integer) **is sel**(x::integer | x >= 0)

Note the use of the select function to filter integer to get values greater than or equal to 0. Most number types are defined in this manner.   ■

7.5 Sequences and Higher-Order Functions

There are two fundamental types of higher-order functions that are useful with respect to sequences. The first includes functions that simply treat the contents of a sequence as a set. Recall from Chapter 5 that a contents function is defined that extracts the contents of a sequence as a set. Given a sequence S, its contents can be extracted as a set using the prefix operation ~S. Thus, it is possible to apply any of the higher-order set functions to sequences. For example, given a sequence S and a boolean expression P, the set of objects from S satisfying P is defined as:

    {**sel**(x::~S | P(x))}

The contents operator extracts the elements of S as a set and the higher-order **sel** function performs the comprehension. Table 7.3 shows how each of the defined higher-order functions for sets can be applied to sequences using the extraction function. Like the previous expression, this function evaluates to the subset of items from S that satisfy P(x).

The second kind of higher-order function treats sequences as sequences generating new sequences from old sequences. Table 7.4 shows the definitions of these sequence functions. Rather than using the set-based higher-order functions, these operations are defined on sequences directly. The two built-in special operations on sequences are image and filter. The image function takes a sequence and an arbitrary function and applies that function to each element in the sequence. To increment the contents of a sequence and maintain the result as a sequence, the map function is applied as:

    image(inc,[1,2,3]) == [2,3,4];

Table 7.4 Special higher-order functions defined over sequences

Operation	Syntax	Meaning
Filter	`filter(P,S)`	*Filter all elements from S that do not satisfy P*
Map	`image(F,S)`	*Apply F to all elements from S and return the resulting sequence*
Fold Left	`reduce(P,i,S)`	*Fold left*
Fold Right	`reduce_tail(P,i,S)`	*Fold right*

Similarly, the `filter` function takes a sequence and removes elements that do not satisfy a predicate. Assuming that the predicate `lt3` exists that is true if its argument is less than three, filtering a sequence for values less than three is achieved by:

```
filter(lt3,[3,1,2,4]) == [1,2];
```

Anonymous functions and **let** forms are particularly useful in conjunction with the image and filter operations. It is unlikely that the `lt3` function just used will ever exist in any library. Using anonymous functions, the filtering operation can be implemented as:

```
filter(<*(x::natural)::boolean is x<3*>,[3,1,2,4]);
```

The use of `filter` is identical in this example, except that the filtering predicate is defined locally and is discarded after the function is simplified and resolved. If a filtering or image function is used repeatedly, the **let** form is useful for defining a local function:

```
let filterFn::<*(x::natural)::boolean*> be
      <*(x::natural)::boolean is x<3*>
   filter(filterFn,[3,1,2,4])
end let;
```

Again the function is identical, but the local function `filterFn` is defined in the **let** form and is used in the filtering activity. Like the anonymous function defined previously, `filterFn` is discarded following the closing of the **let** form's scope. It should be noted that using the **let** form for local function definition in this way can be somewhat cumbersome. If `filterFn` is used repeatedly within the **let** form, or allowing the function to have a name increases readability, then the extra syntax is worthwhile.

EXAMPLE **7.7**

Using Set-Based
Higher-Order Functions on
Sequences

One application of higher-order functions like **exists** and **forall** is a form of comprehension over sequences. Using the contents extraction operation, one can extract values from a sequence and perform comprehension. Assume that we have sequence of `integers`, S, and we would like to determine if all sequence values are positive:

```
allPos(s:sequence(integer))::boolean is
   forall(y::(~S) | x>0)
```

allPos extracts the contents of s to a set and applies the test x>0 to each element returning **true** if all elements are greater than 0. ∎

7.6 Function Inclusion and Composition

Two operations on functions support comparison operations and composition of multiple functions into a single function. *Function inclusion* defines several relations that define when one function is, in effect, a sub-function of another. Specifically, we will define f=<g, f<g, and f=g for any two functions. *Function composition* defines new functions by composing existing functions. The operation (f . g) is defined to represent the function equivalent to the application of g then f to its argument. Specifically, we will define the composition operation so that (f . g)(x)==f(g(x)).

EXAMPLE **7.8**

Using image, filter, and reduce

image and filter are exceptionally useful for manipulating sequences. Using image, defining word functions from bit functions is a simple matter of applying functions to all bits in a sequence. Two functions that perform **xor** and ∗ (and) of one bit across an entire word are:

```
wordXor(b::bit; w::bitvector)::bitvector is
  let helper :: <* (x::bit)::bit *> is
    <* (x::bit)::bit is b xor x *>

wordAnd(b::bit; w::bitvector)::bitvector is image (_*_(b)) w
```

The two definitions represent two different approaches. The wordXor function defines an internal helper function that applies **xor** to the bit input to wordXor and the single input bit x. This function is then applied to each element of the input bitvector. This approach is similar to that used in languages such as Scheme, where currying is not directly supported. The wordAnd function uses currying to define a new function from the binary operation + and the input bit b. This new unary function is applied to each element of the input vector adding the b value with each element. filter is equally useful for extracting values from a sequence or performing searches without resorting to primitive recursion. The following definition is for a function that determines if there are exactly two elements in a bitvector with a value 1:

```
twoOnes(b::bitvector)::boolean is
  #(filter(<*(x::bit)::boolean is x=1*>,b))==2
```

Finally, we can define a parity checking function using reduce over bitvectors:

```
evenParity(b::bitvector)::boolean is %(reduce xor 0 b)
```

The evenParity function starts with 0 and applies **xor** to the accumulated value and the current bit. As each 1 is visited, the accumulated value toggles. The % operation is used to convert the resulting bit to a boolean value. ∎

7.6.1 Function Inclusion

The function type former `<*(d::D)::R*>` defines the set of functions mapping D to R. This set is in all ways a Rosetta type and can be manipulated as a set. Thus, operations such as subset and proper subset are defined over functions. Subset applied to functions is referred to as a *function inclusion* operation. Function containment, `f1=<f2`, holds when one function is fully contained in another function or function type. Assuming `f1(x::d1)::r1` and `f2(x::d2)::r2`:

$$\texttt{f1 =< f2} \overset{\text{def}}{=} \texttt{d2=<d1 and forall(x::d1 | f1(x) == f2(x))}$$

`f1` is contained in `f2` if and only if the domain of `f2` is a subset of the domain of `f1` and for every element of `f2`'s domain, `f1(x)` is equal to `f2(x)`. Exploring function inclusion's several cases reveals where it applies.

The simplest case is when `r1` and `r2` are specified as sets, where the parameter `x` is not involved in the definition. Examining the function inclusion law, the universal quantifier falls out and the following relationship results:

$$\texttt{f1 =< f2} \overset{\text{def}}{=} \texttt{d2 =< d1 and r1 =< r2}$$

In this case, `f1` is included in `f2` when (i) `dom(f2)` is contained in its domain and (ii) its range is contained in `ran(f2)`.

A second case occurs when `r1` is an expression and `r2` is a **set**. Instantiating the function inclusion law results in the following statement:

$$\texttt{f1 =< f2} \overset{\text{def}}{=} \texttt{d2 =< d1 and forall(x::d1 | f1(x) in r2)}$$

`r2` is a constant value independent of `x`. Therefore, the law requires that applying expression `r1` to actual parameter `x` results in an element of `r2`. This is equivalent to the previous result and can be simplified to:

$$\texttt{f1 =< f2} \overset{\text{def}}{=} \texttt{d1 =< d2 and ran(f1) =< r2}$$

As an example, consider the increment function defined over natural numbers. It should hold that:

```
inc :: <*(x::natural)::natural*>
```

Instantiating the function inclusion law gives:

```
1. inc :: <*(x::natural)::natural*>
2.    == natural =< dom(inc) and
         forall(x::natural | inc(x) in natural)
3.    == natural =< natural and
         forall(x::natural | x+1 in natural)
4.    == true and true
```

Thus, inc is of type `<*(x::natural)::natural*>`. It is interesting to note that this is exactly the relationship that must be checked for every definition of the form:

`f(x::`T_d`)::`T_d `is` e`;`

as it indicates that the actual function is of the same type as the specified signature.

The final case defines when one constant function is included in another constant function. In this case, both d1 and d2 are expressions and the most general expression of the function inclusion law must be applied.

First, consider determining if the increment function is included in itself. Clearly, this should be the case and the function inclusion law supports the assertion:

```
1. inc =< inc
2.   == dom(inc) =< dom(inc) and
        forall(x::natural | inc(x) = inc(x))
3.   == natural =< natural and
        forall(x::natural | x+1 = x+1)
4.   == true and true
```

This holds because for any Rosetta item, i=i holds by definition. Consider the case of determining if increment is contained in identity over natural numbers. In this case, the law should not hold:

```
1. inc =< id
2.   == <*(x::natural)::natural is x+1*> =<
          <*(x::natural)::natural is x*>;
3.   == dom(id) =< dom(inc) and
        forall(x::natural | inc(x) = id(x))
4.   == natural =< natural and
        forall(x::natural | x+1 = x)
5.   == true and false
```

false is obtained from the second expression by the counter example provided by x=0 as 0+1 /= 0.

When f1(d::d1) is e1, the following relationship holds:

```
f1 =< <*(x::d2)::r2*> ==
        d2=<d1 and forall(n::d2 | f1(n) =< r2*>)
```

or

```
f1 :: <*(x::d2)::r2*> == d1=<d2 and ran(r1) =< r2
```

The function containment law gives the criteria by which one function may be said to be included within a function type. Each function type defines a set of functions consisting of all those functions included in it. This means that any function can be used as a type, or set, and all the containment laws for sets apply to them. This is particularly useful when using a function that returns a set rather than a single value. Consider the function `<*(n::natural)::natural*>`. This function defines the set of all functions that take a natural number as an argument and return a natural

Table 7.5 Function equivalence and inclusion properties

Operation	Format	Definition
Equal and not equal	f=g, f/=g	f=<g **and** f>=g, -(f=<g) **or** -(f>=g)
Ordering relationships	f>=g, f>g,	g=<f, g=<f **and** f/=g,
	f=<g, f<g	f=<g, f=<g **and** f/=g
Composition	f . g	f(g(x))

number. Rosetta allows the user to ask if a given function is contained in that set and is a member of that type. For example, consider:

```
succ(n::natural)::natural is n+1;
```

We wish to determine:

```
1. succ(n::natural)::natural is n+1 ::
      succ(n::natural)::natural;
2.   ==  (natural=<natural) and
            forall(n::natural | succ(n) in natural)
3.   ==  true and forall(n::natural | (n+1) in natural)
4.   ==  true and true
5.   ==  true
```

Assuming that f(x::df)::rf and g(x::dg)::rg, the operations defined in Table 7.5 are defined over two functions. Functional equivalence checks to determine if every application of f and g to elements from the union of their domains results in the same value. Specifically, f(x) = g(x) for every x in either domain. Function inequality is defined as the negation of function equality.

7.6.2 Function Composition

The function composition operator takes two functions and composes them to form a third. The notation (f · g) represents the composition of functions f and g. The semantics of "g then f" is defined as:

$$(f . g)(x) \stackrel{\text{def}}{=} f(g(x))$$

We define function composition because we can represent the composition without referencing its input parameter or parameter type. This allows writing functions in the *point free* style, a general style useful for writing abstract specifications.

A precondition on the application of function composition g . f is that every value to which g can be reduced must be a legal input to f. This relationship can be specified simply using the ran and ret functions:

```
ret(g) =< dom(f)
```

A simple application of function composition can be used to define a function that adds 2 to its input by composing the increment function with itself:

```
inc(x::integer)::integer is x+1;

plus2 :: <*(x::integer)::integer*> is (inc . inc);
```

The composition is legal because the range of `inc` is the same as its domain. Thus, `ran(inc)=<dom(inc)` holds. A more interesting case deals with an increment function defined over naturals. Changing the definition slightly gives:

```
inc(x::natural)::natural is x+1;

plus2 :: <*(x::natural)::posint*> is (inc . inc);
```

In this case, the range of `inc` is `posint` rather than `natural`. However, `posint=<natural` still holds and the composition is still valid.   ∎

User-Defined Types

Because any Rosetta item whose value is a set can be used to create a type, creating new types is the same process as creating new sets. Rosetta provides three basic mechanisms for forming sets and types: (i) listing the elements explicitly, (ii) filtering or composing existing sets, or (iii) defining functions for constructing the elements of the set. Listing the elements of a set, referred to as *extension*, is the simplest mechanism for defining new types. The new type is formed by simply using the set former to list the elements of the type. Filtering existing sets, referred to as *comprehension*, involves using the **sel** operation or one of its derivatives to extract a set of items from an existing set. Finally, defining functions to construct type elements, referred to as *constructive specification*, involves defining functions that generate all elements of a type.

Types are specified using the expression language to define sets making Rosetta's type system dependent. Specifically, the same language used to define expressions that use declared items is used to define their types. Most programming languages use a distinct language subset for defining types that does not include anything requiring evaluation. Although this restriction makes type checking far simpler, Rosetta uses a dependent type system due to its expressive power.

An excellent example of Rosetta's type system's power is using comprehension to define item properties. Defining a set by comprehension is defining a property that each member of the new set must have. Using Rosetta's set comprehension capabilities, we can define sets that assert a property over all members of a type. This is a simple and powerful way of defining new properties for items that is difficult or impossible to achieve without dependent types.

Because Rosetta types are essentially set values, they are first-class in the language. Types can be passed as actual parameters to facets and functions, and type items can be variable or constant. They can be created, compared, observed, and transformed like any other Rosetta value using the same expression language.

8.1 Defining New Types

When a new Rosetta type is defined, it must either be a subtype of an existing type, or a new base type that has no supertype (Table 8.1). The standard notation for Rosetta declarations applies equally to the definition of new types. Recall that the notation for a item definition is:

names :: *T* ⟦ **is** *e* ⟧ ;

where *names* is a comma-separated list of type names, *T* is the set of values the new type's value is drawn from, and *e* is a Rosetta expression whose value determines the values associated with the new type. Using this notation, the Rosetta type tri can be defined using extension as a subtype of the integer type:

```
tri :: set(integer) is {-1,0,1};
```

In this declaration, the label is tri, the type **set**(integer), and the value {−1,0,1}. The **set** type former creates the powerset of its argument. Thus, tri must be a subset of the integer set. The **is** clause defines the value of tri as the set {−1,0,1}, an element of the powerset of integers and a legal value for this type. Thus, we have defined a new item, tri, that can be used as a type to define both variables and constants:

```
high :: tri is 1;
x :: tri;
```

To distinguish types from sets, Rosetta provides the keyword **subtype** to indicate (i) that a new type is being defined and (ii) the supertype of the new subtype. The **subtype** synonym is semantically the same as **set**, but indicates the intent of a declaration. The previous definition of tri would appear as follows in a definition:

```
tri :: subtype(integer) is {-1,0,1};
```

The most common mechanism for defining new types is using the **subtype** qualifier to to specify the type expression. The notation:

```
x::subtype(integer) is sel(x::integer | x>=0);
```

declares that x is a set of items from integer. It is semantically equivalent to the previous notation. Types defined in this manner are referred to as *interpreted subtypes* because their supertype is known and the value of the new type is also known. In general, new types are defined using the following notation:

name :: **subtype**(*T*) ⟦ **is** *e* ⟧ ;

where *name* is the name of the new type and *T* is the associated supertype. The optional expression, *e*, defines the value of the new type. It is mandatory that the range of *e* is a subset of *T* in the same manner as any other item declaration. In declarations of this sort, *name* is considered subtype of *T* while *T* is the supertype of *name*. Treated as sets, the relationship *name=<T* must hold for all values of *name*.

Like traditional item declarations, the **is** clause is optional, allowing the definition of a subtype whose specific value is not known. For example:

```
time::subtype(integer);
```

defines a new item called time that is a subtype of integer, whose specific value is not known. The difference between this definition and previous definitions is that the value of time is not known. It is known that the values associated with time are found in integer, but those values are otherwise unconstrained. This supports incomplete definitions of types. Specifically, some properties of the type are known, but its specific contents are not. Types declared in this fashion are called *uninterpreted subtypes* because their supertypes are known, but their values are not.

At times it is useful to define a new type that is not a subtype of any existing type. Rosetta does not allow type definitions that do not refer to some kind of supertype in their declaration. To achieve the desired result, the type **top** is used as the supertype. The definition:

```
tri ::subtype(top) is {-1,0,1};
```

defines a new type called tri whose values are −1, 0, and 1 and whose supertype is **top**. In this definition, an item of type tri will not inherit operations from number or element types.

Given the alternative definition:

```
tri ::subtype(integer) is {-1,0,1};
```

operations over integer values are inherited by the new type. The previous definition breaks the inheritance chain forcing the user to define operations over the new type. Types defined in this manner are referred to as *interpreted types*, as their values are known but their supertypes are known to only be **top**.

In both cases existing values are used to populate a new type. When the literal −1 is seen in a specification, some mechanism must be used to determine its associated type. The most common approach is to simply use a type assertion such as −1::tri or −−1::integer to specify the desired type. In some cases, the appropriate type can be inferred. However, it cannot be inferred in the general case.

At times it is desirable to define new types, where both value and supertype are unknown. This definition style defers all properties of the new type to definition using terms within the facet. Such types are defined using **top** as the supertype, but omitting the **is** clause and the explicitly specified value. The definition:

```
tri ::subtype(top);
```

defines a type whose supertype and value are not known. Such types are referred to as *uninterpreted types* or *sorts* in definitions. They are particularly useful at high levels of abstraction where some, but not all, properties are known.

The use of **subtype(top)** is quite common in Rosetta specifications and always indicates the definition of a new type. For readability, the keyword **type** is introduced as a synonym for this construct. The definitions:

```
tri ::type is {-1,0,1};
tri ::type;
```

are identical to the previous definitions, but are easier to read and interpret.

Types need not be defined starting only from elemental values, but can be defined from any type. For example, consider the definition of a type whose values are subsets of a set T:

```
s::subtype(set(T));
```

The notation **set**(T) defines the type containing all possible subsets created from elements of T. The notation x::**set**(T) therefore states that x is a single subset of T. Using the **subtype** form defines s to be an arbitrary collection of subsets of T. The **subtype** form simply says that s may include many sets. This is fundamentally different than the declaration:

```
s::set(set(T));
```

In this case, s is a single set of subsets from T, rather than a set of subsets from T. The notation **set**(T) again defines a type that contains all subsets of T. In this case, the outer set former evaluates to all possible subsets of all possible subsets from T. These types can be further restricted, as in:

```
set4 :: subtype(set(T)) is
   sel(x:: subtype(set(T)) | forall(t::x | #t=4));
```

where set4 is the set of subsets of T that contain exactly four elements. The notation z::set4 declares z to be a singleton element of type set4, or simply a subset of T containing four elements.

One reminder about the distinction between the notations x::T and x::**subtype**(T) as declarations: The first says that the value of x is a single element of type T; the second says that the value of x is a set of values selected from T. If the first definition is used as a type, then only single element types are allowed.

Table 8.1 Forms for defining types classified by interpretability and subtype

	Constant Type	Variable Type
Base Type	T :: **type is** V;	T :: **type**;
Subtype	T :: **subtype**(S) **is** V;	T :: **subtype**(S);

The second explicitly allows sets. In contrast, if these statements are used as terms in a specification, they are identical.

8.2 Defining Types By Extension

There are two mechanisms for defining new types by extension. Using *set formers*, values for a type can be listed explicitly. However, those values must already exist in the Rosetta value space. Using *enumerations*, sets of values are defined where values must be fresh and cannot appear in Rosetta prior to the declaration.

8.2.1 Using Set Formation Operators

The simplest mechanism for defining a type is to simply list its elements or combine existing items to form a set. In Rosetta, the items in a set can be listed explicitly by using the set former. For example, the definition of type bit has the following form:

```
bit ::subtype(natural) is {0,1};
```

Because {0,1} is an element of **subtype**(natural), this represents a legal type definition. Further, the new bit type inherits operations from the natural numbers. Other set formation operations, such as intersection and difference, can be used to form types in the same manner as for the set former. They are used far less often, but still remain useful. For example, the type containing only positive integers, posint, can be defined as:

```
posint :: subtype(natural) is sel(x::natural | x>0);
```

using the selection function to filter out all values greater than 0.

As noted earlier, there is some danger in defining new types whose values are shared with other existing types. When processing specifications, Rosetta tools must determine the type of −1, 0, and 1. If the new bit type is defined without reference to integer, the only type that can be inferred is **top**. To assist tools, type ascription can be used to indicate the type explicitly:

```
1::tri ;
```

identifies 1 as the value from tri, while:

```
1::integer;
```

identifies 1 as the value from integer. With two types associated with 1 and no common supertype other than **top**, this annotation is required in virtually all situations. Using an enumeration to define new types like this will avoid this problem.

8.2.2 Enumerated Types

Enumerations are types whose values can be specified by explicitly listing them. Unlike definitions using set enumeration, the values of an enumerated type cannot be defined previously and are subtypes of element only. Defining an enumerated type is achieved by declaring the type and defining a collection of values associated with it using the **enumeration** former:

```
colors::type is enumeration[red,yellow,blue];
```

The notation yellow defines a new item whose lexical representation is yellow and whose value is also yellow. Thus, yellow becomes a new value in the current value space. Enumerations in Rosetta differ from enumerations in traditional programming languages, as the new type definition does not imply an ordering on its elements. Ordering must be introduced separately if needed. Any variable whose type is an enumerated type is an element and thus a scalar.

Because enumeration values are fresh and cannot exist in the Rosetta type space prior to the enumeration declaration, the problem with ambiguous types for values is avoided. Ascription can be used in the same manner as described previously, but it need not be, as each enumeration value is associated with exactly one type. It is possible to define subtypes consisting of enumeration values. This reintroduces the need for ascription.

The semantics for enumerated types is defined by elaborating the type declaration to a constructed type. Described later in this chapter, constructed types define a type by specifying a collection of constructors for all elements of that type. An enumeration is simply a shorthand for defining constructed types with constant constructors.

Using enumerated types, we can define a three-valued logic that adds an unknown value to the normal logical high and low values. The declaration of this type using enumerated types is:

EXAMPLE 8.1

Defining and Using a Three-Valued Logic Using Enumerations

```
tri :: type is enumeration[high,low,unknown];
```

We can now define standard operations over this type using high, low, and unknown:

```
tri_and(x,y::tri )::  tri is
case x of
  {high} -> y
  | {low} -> low
  | {unknown} -> if y=low then low else unknown end if
end case;

tri_or (x,y:: tri)::  tri is
case x of
  {high} -> high
  | {low} -> y
  | {unknown} -> if y=high then high else unknown end if
end case;
```

`tri_and` and `try_or` define the logic of conjunction and disjunction for this three-valued logic. Other functions can be defined similarly. Using these definitions, we can define facet models for **and** and **or** gates, respectively:

```
facet tri_and_gate(x,y::input tri ; z::output tri)::state_based is
begin
   update: z' = tri_and(x,y)
end facet tri_and_gate;


facet tri_or_gate(x,y::input tri; z::output tri)::state_based is
begin
   update: z' = tri_or(x,y);
end facet tri_and_gate;
```

8.3 Defining Types By Comprehension

Defining types by comprehension involves filtering or transforming existing types to create new subtypes. Using the **sel** function or `filter` functions defines new types by filtering values from old types. Using the `ran` or `image` functions defines new types by transforming old types.

8.3.1 Using the Selection Function

An excellent example of defining types by comprehension is the definition of the built-in type `natural`. Ideally, `natural` should include the `integer` values that are greater than or equal to zero. Using this fact, the standard definition for natural numbers is:

```
natural :: subtype(integer) is sel(x::integer | x >= 0);
```

In this type definition, the comprehension operator **sel** chooses all elements of the type `integer` that satisfy the relationship $x >= 0$. Because `natural` is a subtype of `integer`, all operations defined on `integer` are also defined on `natural`.

Similarly, the **sel** operator can be used to form a type called `byte` consisting of 8-bit bitvectors . Here the selection operator will be used again, with the length of the bitvector being checked:

```
byte :: subtype(bitvector) is sel(b::bitvector | #b=8);
```

In this definition, the selection operator chooses elements from the `bitvector` type that are of length 8.

8.3.2 Using the Range Function

While the selection operator chooses elements from a type, the range operation transforms each element of a type. An example of using `ran` to define a new type is a potential definition of the even integers. Instead of using the comprehension operator, here the ran operation is used as an image function:

```
even :: subtype(integer) is ran(x::integer | 2*x);
```

Here the even numbers are generated by multiplying each integer value by 2.

As demonstrated here, the selection and range operations are excellent choices for defining new types by comprehension. Other higher-order functions, such as **forall** and **exists**, are far less useful, as they evaluate to single boolean values.

8.3.3 Sets as Types — A Caution

As noted earlier, any Rosetta set can be used as the value of a type. This has numerous advantages, not the least of which is clarity. However, it quite easy to misuse this capability to develop specifications that are exceptionally difficult to analyze. Using operations such as set union, intersection, and difference to form new types can introduce specification complexities that make analysis virtually impossible. We have already seen the need to use type annotations to disambiguate literals shared between types. New types formed from arbitrary sets must be used carefully, with full understanding of their purpose and how they impact analysis.

8.4 Defining Constructed Types

The constructed type definition syntax provides a mechanism for defining *constructor*, *observer*, and *recognizer* functions in a single definitional notation. Constructor functions create values associated with the new type. These values behave as any other values and are treated as normal forms that are not evaluated. Together, the collection of constructor functions can create every value associated with the type. The observer functions define properties associated with values of the new type. Observer functions specify and calculate properties for constructed type values. Finally, recognizer functions are special predicates that indicate the constructor used to create a value. Summarizing: (i) constructor functions create new values, (ii) observer functions define properties, and (iii) recognizer functions indicate how the value was created.

As an example of constructed type definition, consider the following definition for a binary tree of integers:

```
intTree :: type is data
  nil::empty |
  node(L::intTree,v:integer,R::intTree )::nonempty;
end data;
```

The item intTree is defined to be a new, fresh type using the **type** indicator. This declaration simply says that the new type's supertype is **subtype(top)**.

The **is** clause is used to give the type a value and the **data** keyword is used to indicate that the value of this type will be given a value.

The constructed type definition provides two constructors for intTree: (i) the constructor function nil and (ii) the constructor function node. The nil function creates an empty tree and is recognized by the boolean function empty. The recognizer function is defined so that if its argument is the nil constructor function, it will return true, otherwise it returns false. The node constructor creates a nonempty tree from a value and a left and right subtree. Note that the type is recursive in that intTree is referenced in the definition of node. The nonempty recognizer is true whenever its argument was created by a call to the node function. In addition, the observer functions L, v, and R observe parameters of the node constructor.

A tree with one node whose value is 0 is defined using the node constructor:

```
node(nil,0,nil);
```

The recognizers empty and nonempty indicate the constructor used to generate a tree. Specifically, empty is **true** if its argument is nil and nonempty is **true** if its argument is an instantiation of the node constructor. The recognizers nonempty and empty evaluate to:

```
nonempty(node(nil,0,nil)) == true
empty(node(nil,0,nil)) == false
```

A balanced tree with 0 as the root, and 1 and 2 as the left and right nodes, respectively, is constructed as follows:

```
node(node(nil,1,nil),0,node(nil,2,nil ));
```

Here, the node constructor is used to create nonempty left and right subtrees.

Parameter names are used to generate observer functions that return actual parameters from constructor functions. These functions return the actual parameter instantiation of their associated formal parameter. For example, when evaluated on the node constructor, the observers L, R, and v evaluate to:

```
L(node(nil,0,nil )) == nil
R(node(nil,0,nil )) == nil
v(node(nil,0,nil )) == 0
```

In this sense, the constructed value behaves much like a record— so much so that Rosetta uses constructed types in lieu of traditional record structures.

The primary purposes for constructors, observers, and recognizers should be clear from these examples. Constructors create values, observers make observations over values, and recognizers partition the value set. In **if** and other conditional statements, the recognizers are used to guard the use of observers. For example:

```
if nonempty(t) then L(t) else nil end if
```

determines if t is empty and accesses the left subtree if it is not. If t were empty, then L would be undefined.

Constructed type definitions have the form:

```
T :: type is data
    c(parameters)::r
    ⟦ | c(parameters)::r ⟧*
end data;
```

where T names the new type and each subsequent *constructor* definition specifies a constructor name, a collection of typed parameters, and a recognizer name. The format of each constructor declaration is identical to a function definition. The parameters identify the names and types of observers, and the recognizer names a predicate that is true when its argument is constructed with the associated constructor.

Like any other function, constructor functions can be curried. If this is the case, then the results of applying observer functions associated with uninstantiated parameters are not defined. However, constructor functions are not evaluated in the traditional fashion. When a constructor function is fully instantiated, it becomes a value in the Rosetta value space. Thus, it cannot be further evaluated.

The ranges of constructor functions are disjoint, implying that each element of the constructed type is created by exactly one constructor. This implies that each value associated with a constructed type can have only one form. Furthermore, each constructor name must be unique. This allows Rosetta tools to automatically determine what type a constructed value is associated with.

EXAMPLE **8.2**

Defining Records using constructed types

In Rosetta, no special syntax for defining records is defined, as record structures follow directly from constructed types. A record type is a constructed type with a single constructor function that associates values with parameters used as field names. A typical record type will be defined with the following constructive technique:

```
record::type is data
    recordFormer(f_0::T_0 | f_1::T_1 | ... f_n::T_n)::recordp;
end data;
```

where recordFormer is the single constructor, f_0 through f_n are the names of the various fields, and T_0 through T_n are the types associated with those fields. The recognizer recordp is also defined, but is largely unused, as there is only one constructor. Any constructed type definition providing only a single constructor is referred to as a record. To define a specific record type that represents Cartesian coordinates, the following notation is used:

```
cartesian::type is data
    cartFormer(x::real, y::real, z::real)::cartp;
end data;
```

To define an item of this type, the standard Rosetta declaration syntax is used:

```
c :: cartesian;
```

Values can be associated with record items using the canonical **is** form:

```
origin :: cartesian is cartFormer(0,0,0);
```

Accessing individual fields of the record is achieved by applying one of the observer functions associated with a field name. To access field y in the record c, the following notation is used:

```
y(c)
```

Defining a coordinate in Cartesian space using this definition is achieved by:

```
cartFormer(1,0,0);
```

Accessing elements of the constructed coordinate is achieved using observer functions:

```
x(cartFormer(1,0,0))==1;
y(cartFormer(1,0,0))==0;
z(cartFormer(1,0,0))==0;
```

The following notation creates a Cartesian coordinate whose x and y values are known, but whose z value is not specified:

```
cartFormer(1,0);
```

cartFormer(1,0) produces a function of one parameter that returns the completed Cartesian value. Should the function z(cartFormer(1,0)) be specified, it should not pass type checking because cartFormer(1,0) is a function, not a Cartesian value.   ■

EXAMPLE **8.3**

Semantics of Enumerations

Similar to records, it is quite easy to define enumerations using constructed types. The enumeration declaration:

```
colors :: type is [red,yellow,blue]
```

is semantically equivalent to:

```
colors :: type is data
            red()::redp
          | yellow()::yellowp
          | blue()::bluep;
    end data;
```

The constructed type defines three nullary constructors that are now values in the value space. Further, three recognizers are provided that specify what constructor produces a value. In this case, the recognizers are not particularly useful because the equals relationship performs the same function. Specifically:

```
redp(x) == x=red()
```

In a sense, an enumeration is the opposite of a record. A record is defined using one constructor with parameters to indicate record elements. An enumeration is defined using many constructors with no parameters to define different elements.   ■

8.5 Functions as Type Definition Tools

Using functions to define types allows the introduction of parameters in type definitions. Because types are first-class items in Rosetta, any function returning a set can be used to define a Rosetta parameterized type. Consider the following function definition:

```
word(n::natural)::subtype(bitvector) is
  sel(b::bitvector | #b = n);
```

The function signature defines a mapping from natural to a subtype of the built-in bitvector type. The subtype is defined by the **sel** operation to be those bitvectors whose lengths are equal to the parameter n. Thus, word evaluates to the set of bitvectors of length equal to its parameter. We can now use word as a type definition construct. The notation:

```
reg::word(8);
```

defines reg to be a bitvector of length 8. The notation:

```
bv8::subtype(bitvector) is word(8);
```

defines bv8 to be the type containing all bitvectors of length 8.

When using functions as type definition tools it is important to understand that they, themselves, are not types. They generate types, but they are not usable as types prior to evaluation. The word type is an excellent example. Altough word(8) is a type, word is simply a function making the declaration w::word meaningless.

EXAMPLE 8.4

Defining Generic Trees

The intTree example demonstrates a capability for defining a tree containing integer values that can easily be modified to define a tree containing other types. Simply replace the integer type with the item type to be contained in the tree. Using functions and constructed types together, a general tree type definition function can be developed that allows definition of trees containing arbitrary types.

The modified tree definition has the following form:

```
tree(T::type) :: type is data
  nil ::empty |
  node(L::tree(T),v:T,R::tree(T ))::nonempty;
end data;
```

This definition uses the standard function definition format, specifying a function signature and an expression. In this case, tree is defined to take a single, arbitrary type and return another type. The value of the returned type is obtained by evaluating the function's associated expression. In this case, the **data** keyword indicates a constructed type definition. If the tree definition is instantiated with a type, the result is a tree **data** type that contains the instantiated type. For example:

```
intTree::type is tree(integer);
```

is semantically equal to the intTree provided earlier in this section. If the parameter T is replaced by integer in the expression, the same type definition results. The advantage here is that the new tree function can easily and quickly be used to define new tree types with new contents. The tree function is not a type, but simply a function whose result value is a type value. Unfortunately, it is not possible to define a new item t::tree. The tree function must be fully instantiated.

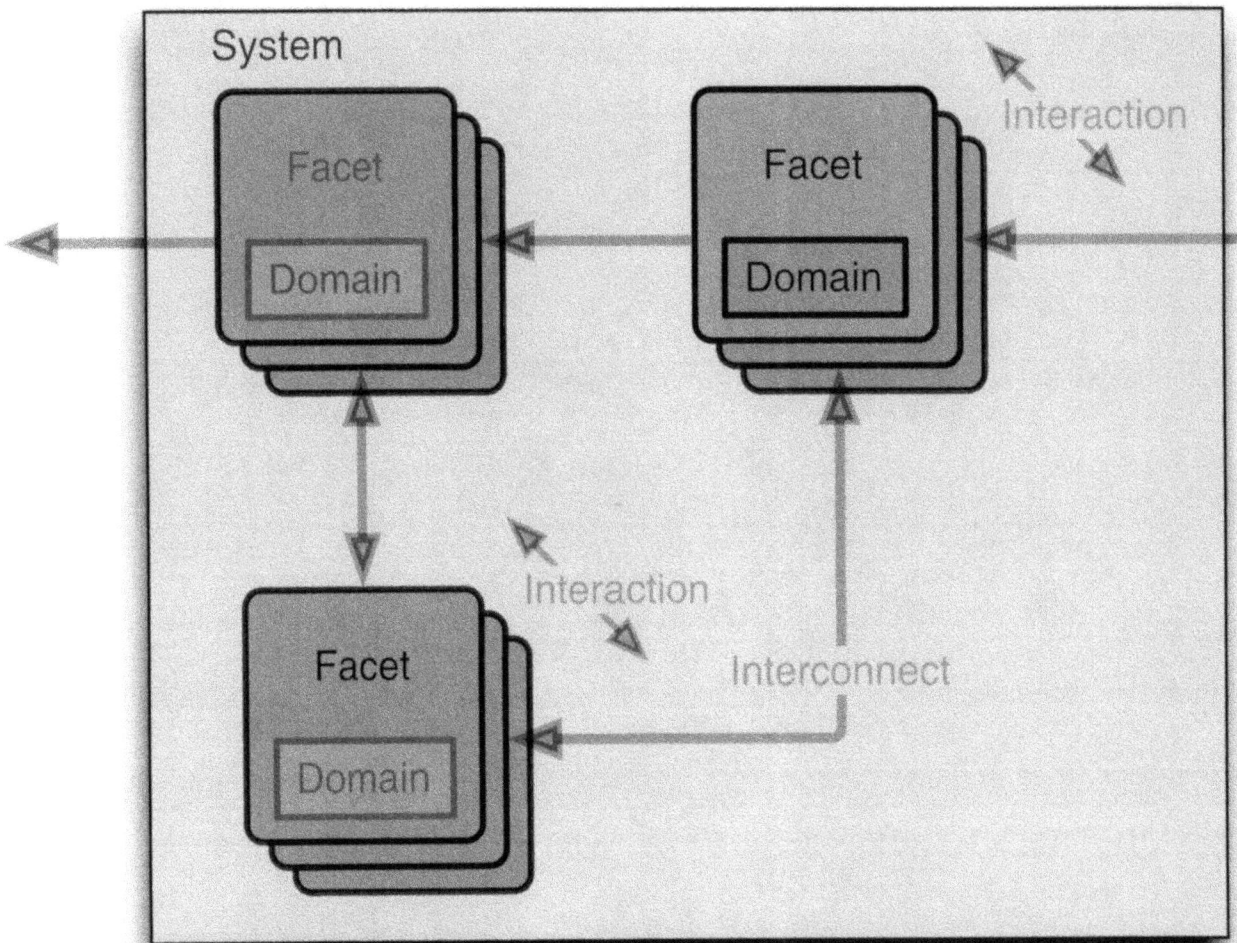

The Facet Language

The basic unit of Rosetta specification is the *facet*. Each facet describes one domain-specific system model. Each facet extends a domain, called its type, that provides a domain-specific modeling vocabulary and semantic basis. Facets are parameterized and may define local items. Facet properties are defined by a collection of terms that are either Boolean or facet valued. Boolean terms define properties directly while facet terms define structure by instantiating and interconnecting facets.

Part III describes the Rosetta the *facet language* used to encapsulate items in structures that define system models. The facet language is used to define and assign domains to facets, define packages containing specifications and libraries where they are stored, and components that encapsulate functional requirements, usage assumptions and correctness conditions.

After completing the chapters in Part III, you will understand how to declare facets, compose facets into structural specifications, define and use components, define libraries, and group specifications together into packages.

Facet Basics

Thus far, our exploration of Rosetta has concentrated on the Rosetta expression language used to define Rosetta items and state mathematical properties over those items. Although the expression language is useful in its own right, it does not provide a mechanism for defining models where multiple properties hold simultaneously over multiple system observers. Nor does it provide mechanisms for modeling how properties change over time.

We look now at the *facet language* for defining the fundamental Rosetta modeling construct used to define system and component models. A *facet* is a parameterized, declarative structure that defines the properties of a system. Each facet definition consists of four major elements: (i) a domain, (ii) parameters, (iii) local declarations, and (iv) terms. A facet's domain identifies the vocabulary and model of computation used as a basis for the facet model. Facet parameters define a model's visible interface and provide a mechanism for customization and instantiation. Facet declarations provide local items defining state and local functions. Facet terms declaratively define system properties by defining properties of parameters and local variables. By defining facets, specifiers define system properties from a particular perspective. By combining facets, specifiers define multiple perspectives for a given system.

Facets differ from constructs in traditional modeling and programming languages in that they use declarative constructs, and the underlying model of computation varies from model to model. Rather than defining a program or executable model exhibiting system properties, facets define those properties directly. This allows Rosetta to be far more expressive and general as compared to an executable specification language or programming language. Abstract properties are easily defined without overspecifying implementation details. Incomplete specifications are allowed and can be refined during the design process. Rather than infer properties by observing system execution, properties are defined directly.

The domain identified by a facet specifies its vocabulary and computational model. Where a state-based model may be appropriate for one facet, a continuous time or frequency domain model may be appropriate for another. Rather than force a single modeling semantics on every facet, Rosetta allows the user to choose vocabulary and semantics for each model individually. Later we will see how these heterogeneous models are composed to define complete systems.

9.1 A First Model—An AM Modulator

The facet am_mod constructed in this section defines a simple model for an AM modulator that exhibits the basic features of a facet model. The AM modulator is a continuous time model that outputs an input baseband signal multiplied by a sinusoidal carrier whose amplitude and frequency are specified as input parameters. This is a classical model of AM modulation, typically expressed mathematically as:

$$s(t) = f(t)\cos(\omega_c t)$$

where $s(t)$ is the output signal, $f(t)$ is the signal, and the cos term is the carrier.

A facet embodying a modulator that implements a system exhibiting these properties must define a *signature, domain,* and *terms.* The signature defines the parameters and local items needed to define the system and consists of a facet name, parameters, and declarations. The domain identifies the underlying computation model used to describe the system. Terms define properties over the items defined in the signature and domain. Defining a facet thus becomes instantiating a standard syntactic template identifying its components:

```
facet F( ⟦ parameters ⟧ ) :: domain is
  ⟦ declarations ⟧
begin
  ⟦ terms ⟧
end facet F ;
```

Defining a facet begins with its interface. When defining an interface, we must decide what quantities associated with the facet will be visible. Specifically, we must identify inputs, outputs, and design parameters. Inputs and outputs represent quantities input and output through parameters at the facet interface. Design parameters define static parameters used to configure a component. The am_mod component will minimally input a baseband signal, $f(t)$, and output a modulated signal, $s(t)$. To make this model customizable, we will also add a design parameter specifying the carrier frequency. Because the modulator has no memory, no local variables are needed.

The facet's *name, parameters,* and *declarations* together define its signature. The parameter list defines inputs, outputs, and design parameters while the declarative region contains local declarations. Beginning the facet declaration we have:

```
facet am_mod(f::input real; s::output real;
             w::design real) :: domain is
begin
  ⟦ terms ⟧
end facet am_mod;
```

We have named our facet am_mod and defined three parameters corresponding to quantities in the earlier equation. Dependent variables become outputs while independent variables become inputs. Each parameter is named with an associated direction and type. Specifically, f is the input signal, s is the output signal, and w is the carrier frequency. All are taken directly from the specification equation. There are no local definitions, thus the declarative region contains no declarations.

The selection of a domain dictates how and when observations of system state are made. As presented thus far, a variable has a label, type, and potential value. When observing a system over time, observable quantities, such as a system's state and parameters, have a trajectory. As they change, variable values can be traced and sequences of values observed. This is the basis of most simulation and model-checking analysis tools. The domain specifies when observations can be made and how those observations are sequenced. Informally, if a variable's value is graphed using a Cartesian coordinate system, the variable value forms the Y axis while the domain specifies the type of the X axis. For example, in the discrete_time domain, quantities are observed at discrete time intervals and the X axis is a discrete time line. In contrast, in the continuous_time domain, quantities are observed continuously over time and the X axis is a continuous time line.

In the original equation, the quantities f and s are functions of a variable t. In the signal processing domain, t typically represents continuous time. Specifically, the quantities $f(t)$ and $s(t)$ can be observed with respect to any continuous time value. To model this in the am_mod facet, we select the continuous_time domain. In this domain, each variable is a mapping from the continuous time line to a value. Thus, any parameter or variable can be observed with respect to a particular instant in time by saying v@t, where v is an item name and t is a time value. If v is specified without t, then the current time is assumed. We will explore this concept further in subsequent chapters and define a general structure for defining domains. For now, it is sufficient to understand that the selection of a domain defines the temporal properties of observations. Continuing to define the am_mod facet, the domain can now be specified:

```
facet am_mod(f::input real; s::output real;
             w::design real) :: continuous_time is
begin
  ⟦ terms ⟧
end facet am_mod;
```

The remaining task is defining properties over quantities that describe the modulator. The modulation equation gives us precisely the information we need for the definition. By choosing the continuous_time domain, the parameters f and s correspond with the functions $f(t)$ and $s(t)$, respectively. The continuous_time domain provides a variable t that corresponds to the current time value. The single term labeled mixer in the completed facet below defines a single system property specifying a relationship between input and output parameters. Specifically, the output parameter, s, is equal to the input signal, f, times a carrier:

```
facet am_mod(f::input real; s:::output real;
             w::design real) :: continuous_time is
begin
   mixer: s=f*cos(w*t);
end facet am_mod;
```

The notation s=f*cos(w*t) specifies that s is equal to f*cos(w*t). The term constrains the value of s by making it equal to a defined quantity. This is not an assignment any more than the equality assertion in the mathematical equation is assignment. The equality operator defines an equivalence relationship between the variable and the quantity.

Other properties can be added to define constraints and additional functionality within the facet. If we would like to specify that the output voltage of the device modeled by the facet should not be greater than 5 volts, we simply add a term indicating this constraint:

```
facet am_mod(f::input real; s:::output real;
             w::design real) :: continuous_time is
begin
   mixer: s=f*cos(w*t);
   out_limit: abs(s) =< 5;
   in_limit: abs(f) =< 10e-3;
end facet am_mod;
```

The term out_limit asserts that the absolute value of the output signal, s, must always be between -5 and 5 volts. The term in_limit asserts that the input voltage can never exceed 10 mV. With the addition of these terms, properties defining both behavior and constraint information are present in the specifications.

This facet definition process demonstrates several important characteristics of Rosetta definitions and facets. First, the specification is declarative. Each term defines a property that must hold over a collection of items, not an executable statement. Second, a domain is used to define how the system will be defined and observed. Traditional specification and simulation languages use a single domain or a fixed collection of domains. As we shall see later, Rosetta provides a variety of domains and allows users to define their own. Finally, a system specification parallels the style used in the application domain. The mixer equation is virtually identical to the mathematical formula, making the Rosetta specification easy to read and write. As we learn more about domains, we shall see how they contribute to this Rosetta capability.

9.2 Composing Models — Adding Constraints

Other domains define computation differently, providing Rosetta with a heterogeneous definition capability. We can modify the AM modulator definition to separate constraints and functional description into two specifications:

```
facet am_mod_fn(f::input real; s::output real;
                w::design real)::continuous_time is
begin
   mixer: s=f*cos(w*t);
end facet am_mod;


facet am_mod_const(f::input real; s::output real)::static is

begin
   out_limit : abs(s) =< 5;
   in_limit : abs(s) =< 10e-3;
end facet am_mod;

am_mod(f::input real; s::output real; w::design real) :: static is
   am_mod_fn(f,s,w) * am_mod_const(f,s);
```

The am_mod_fn facet is identical to the original facet defining the AM modulator. The am_mod_const facet contains only the constraints specified for the AM modulator in the second mode. In this case, the domain is static, reflecting that the constraints are invariant — specifically, that regardless of any concept of time, the magnitude of the input and output voltages must never exceed 10 mV and 5 V, respectively. The final definition specifies a new version of the am_mod specification that is the product of the AM modulator's functional specification and associated constraints. The result is a new model that reflects properties specified in both original models. The domain of this model is static, reflecting the fact that this is the only common domain available for this composition.

This example reflects the compositional nature of Rosetta specifications. Functional requirements and constraints are represented in separate facets using separate semantic domains. The facet product defines a new model that must exhibit both the specified functional behavior and operational constraints. The concept of facet product is exceptionally useful when describing heterogeneous system aspects and will be discussed in detail in Chapter 15.

9.3 Combinational Circuits — A Simple Adder

We have seen facets define continuous systems and static constraints. State-based specifications, like digital components, are just as easily described using the same techniques and constructs. The following equations define digital equations for a full adder:

$$z = x \oplus y \oplus c_{in}$$

$$c_{out} = (xy + xc_{in} + yc_{in})$$

Using identical techniques, we can define a facet representing the function of the full adder. As before, we start with the standard facet template:

```
facet F( ⟦ parameters ⟧ ) :: domain is
   ⟦ declarations ⟧
begin
   ⟦ terms ⟧
end facet F;
```

We will call the model `adder_fcn` to reflect that it specifies the basic function of an adder. From the preceding equations, we identify z and co as outputs, with x, y, and ci as inputs. Instantiating the facet template with this information results in:

```
facet adder_fcn(x,y,ci::input bit;
                z,co::output bit) :: domain is
begin
   ⟦ terms ⟧
end facet adder_fcn;
```

We must now select an underlying computation model and instantiate the domain. The `continuous_time` model used for the AM modulator can certainly be used here. In exactly the same manner, the preceding equations can be used as terms in the model definition:

```
facet adder_fcn(x,y,ci::input bit ;
                z,co::output bit) :: continuous_time is
begin
   sum: z = x xor y xor ci;
   carry: co = (x and y) or (x and ci) or (y and ci);
end facet adder_fcn;
```

In this definition, values are specified for both the sum and carry values. The mathematical expressions are translated into Rosetta syntax, but the semantics remains the same. The terms state that at any time, the sum and carry outputs are equal to the values specified by their respective terms.

The first specification of the adder's function is ideal. The relationship between the output parameters and their equations holds constantly for any time. We know that such circuits do not exist — there is always some delay in the circuit. Ideally, we should be able to say that the defined relationships hold between current values of inputs and output values sometime in the future. Because the `continuous_time` domain makes time explicit, defining such a specification is relatively simple:

```
facet adder_fcn(x,y,ci::input bit ;
                z,co::output bit) :: continuous_time is
begin
   sum: z@(t+5e-6) = x xor y xor ci ;
   carry: co@(t+6e-6) = (x and y) or (x and ci) or (y and ci);
end facet adder_fcn;
```

The difference is the use of the @ operator to specify the value of a variable at some time in the future or past. Understanding the new terms is a simple matter of reading the specification as "z at time t+5e-6 is equal to ...". What the term says is that at some time 5 microseconds in the future, the value of z is specified by the equation. Note that when the "@" operator is omitted, the current time or state is assumed.

So, 5 microseconds in the future, the value of z will be equal to x **xor** y **xor** z for all values in the current state. The delay through the circuit differs for the two outputs, but is explicitly specified in the definition.

If desired, we can use a design parameter to allow reuse of the facet specification:

```
facet adder_fcn(x,y,ci::input bit ;
                z,co::output bit)
                delay :: continuous_time is
begin
   sum: z@(t+delay) = x xor y xor ci;
   carry: co@(t+delay) = (x and y) or (x and ci) or (y and ci);
end facet adder_fcn;
```

When the facet is instantiated, the actual parameter associated with delay specifies the delay time through the circuit.

An alternate definition of the counter uses the state_based domain to allow for non-ideal behavior without choosing a specific time model:

```
facet adder_fcn(x,y,ci::input bit ;
                z,co::output bit) :: state_based is
begin
   sum: z' = x xor y xor ci;
   carry: co' = (x and y) or (x and ci) or (y and ci);
end facet adder_fcn;
```

The delay parameter is gone along with references to actual time values or increments in the specification. The notations z' and co' reference values in the next state, but do not indicate how the next state is obtained or how it is observed. The specification simply states that there is a next state, acknowledging that instantaneous change does not occur.

The advantage of this specification is its abstract nature. Time is held abstract, allowing function to be specified before details of time are known. This is precisely how most digital designers define initial system specifications.

9.4 Defining State — A 2-bit Counter

Thus far, each defined facet is a stateless mapping of inputs to outputs requiring no representation of internal state. To demonstrate the use of internal state, we will now define requirements for a simple 2-bit counter with a single reset input. The equations for such a device are:

$$s' = (s + 1) \bmod 4$$

$$o = s$$

Where s is the state, s' is the next state, and o is the output.

Again, we start with the same facet template:

```
facet F ( ⟦ parameters ⟧ ) :: domain is
    ⟦ declarations ⟧
begin
    ⟦ terms ⟧
end facet F;
```

We select a name, `counter`, and define reset and clock inputs, and a value output. We use the parameterized `word` type to define a 2-bit value:

```
facet counter(reset::input bit ;
            value::output word(2);
            clk::input bit) :: domain is
    ⟦ declarations ⟧
begin
    ⟦ terms ⟧
end facet counter;
```

The `counter` differs from previous definitions. First, it is stateful — the value of the output is dependent on inputs and on the previous output value. Thus, we need to define an internal state for the mode. Second, it is a finite state system — a 2-bit counter has exactly four states. Based on these observations, we can select a domain and define an internal state.

Remember that a domain defines the ordering of computations. Because state and order are intimately tied, the domain and state definition are also dependent. Rosetta provides a special domain called `finite_state` for finite state systems. This domain takes a single parameter that defines the type of the state. If that type is finite, then the state set is finite. For our counter, the state set is defined by all 2-bit values, precisely the set provided by `word(2)`. This observation allows us to define the facet's domain:

```
facet counter(reset::input bit; value::output word(2);
            clk::input bit) :: finite_state(word(2)) is
    ⟦ declarations ⟧
begin
    ⟦ terms ⟧
end facet counter;
```

In the same way that `continuous_time` provided a current time value, the `finite_state` domain provides a current state value named s and a next state function that maps one state to another, `next(s::state)::state`. Our task is to provide terms defining these quantities from the earlier equations.

The facet in Figure 9.1 defines the complete counter model. The `next_state` term defines the next state function for the counter. In the state-based domains, the tick decoration indicates the value of a symbol in the next state. Using this notation, the `next_state` term declares that s' is state 00 if `reset` is high, else it is s **mod 4** if the clock is rising. Two local functions are defined to calculate the next state in binary and identify a rising edge. The next state function is a direct application of a **case** statement defining the next state by extension. The function `rising` uses the built-in predicate `event(x)` to detect a change in the value of its parameter and the equivalence x=1 to determine if the parameter is high.

```
facet counter(reset::input bit ; value::output word(2);
              clk::input bit )::finite_state (word(2)) is
  inc_mod4(x::word(2))::word(2) is
    case x is
      {b"00"} -> b"01" |
      {b"01"} -> b"10" |
      {b"10"} -> b"11" |
      {b"11"} -> b"00"
    end case;

  rising(s::bit )::boolean is event(s) and s=1;

begin
  next_state: s' = if reset=1
                     then b"00"
                     else if rising(clk)
                            then inc_mod4(s)
                            else s
                          end if;
                   end if;
  output: value = s;
end facet counter;
```

Figure 9.1   A 2-bit counter with reset.

9.5   Defining Structure—A 2-bit Adder

In addition to directly defining system behaviors, terms can describe the structure of a system using facet instantiation and renaming. The following definition constructs a 2-bit adder from the full adder defined previously:

```
facet adder2_fcn(x0,x1,y0,y1::input bit, co::output bit,
                 z0,z1::output bit)::continuous_time is
  ci :: bit;
begin
  b0: adder_fcn(x0,y0,0,z0,ci);
  b1: adder_fcn(x1,y1,ci,z1,co);
end facet adder2_fcn
```

The 2-bit adder functional model is a structural model and differs substantially from the monolithic, 1-bit adder model. Instead of defining facet properties directly, the 2-bit adder is defined structurally by instantiating and interconnecting facet models. Two 1-bit adders are used in a notation that is quite similar to structural VHDL. In this case, the terms b0 and b1 are facets representing 1-bit adders rather than boolean values.

The 1-bit adder facets communicate and interact with their environment through parameter instantiation. The internal variable ci defines the internal carry signal that communicates carry out from the first adder to carry in from the second. The facets interact with the external environment by instantiating parameters with system inputs and outputs. The adder is instantiated with the first bit of the 2-bit input while the second adder is instantiated with the second.

It is important to note that the adder models within the 2-bit adder are unique instances of the 1-bit adder. A technique called *relabeling* is used to create fresh copies of the 1-bit adder component within the 2-bit adder. Reference to the 1-bit adder is maintained to ensure that when the 1-bit adder model changes, the 2-bit adder will change to reflect the new model.

9.6 Specification Reuse — Using Packages

The Rosetta **use** clause allows specifiers to include packaged definitions in a facet. A package is a special facet that contains declarations. The syntax:

> **use** *name*;

immediately preceding a facet declaration includes the items exported from *name* in the scope of the facet. For example, the following facet uses two packages to define a simple instruction interpreter model for a CPU:

```
use cpu_utils(8);
use clk_utils;
facet instruction_interpreter
   (clk::input bit; datain:: input word(8);
    address,dataout::output word(8))::state_based is
  registers :: cpu_utils.regfile;
begin
  if clk_utils.rising(clk)
     then ...
     else ...
  end if;
end facet instruction_interpreter ;
```

The **use** clauses import definitions from cpu_utils and clk_utils, respectively. Within the facet, package instances are included in the declarations and terms section. The term fragment uses the definition of rising to monitor the clock input for a rising edge. The dot notation, clk_utils.rising, is used to make certain the definition from package clk_utils is used. If no other definition of rising is provided in the scope of the term, the qualifier may safely be omitted.

The declaration of registers uses the type cpu_utils.regfile to define a register file for the definition. Note that the **use** clause including cpu_utils has an actual parameter that specifies the size of the structures defined in the package. In this case, the parameter defines 8-bit devices. Packages are parameterized like facets except that all parameters must be of kind **design**, implying that they do not change with respect to state or time.

Using package parameters, we can parameterize the word size of the entire instruction interpreter model:

```
package instruction_interpreter_pkg
   (width::natural)::state_based
use cpu_utils(width);
use clk_utils;
facet instruction_interpreter
   (clk::input bit; datain:: input word(width);
    address,dataout::output word(width))::state_based is
  registers :: cpu_utils.regfile;
begin
  if clk_utils.rising(clk)
     then ...
     else ...
  end if;
end facet instruction_interpreter;

end package instruction_interpreter_pkg;
```

The package `instruction_interpreter_pkg` defines a single parameter, `width`, used to specify the word width of the model. The **use** clause:

```
use instruction_interpreter_pkg (8);
```

instantiates the package parameter so that we have the same model as the original definition.

Packages, components, and other library constructs are discussed fully in Chapter 11. Numerous standard packages are defined for Rosetta that allow functions ranging from simple mathematical definitions through manipulating Rosetta constructs using reflective operations.

9.7 Abstract Specification — Architecture Definition

One area where abstract specification is important is definition of general-purpose architectures. Such architectures are defined using facets with facet-type parameters. This advanced definition demonstrates numerous Rosetta capabilities that will be defined in subsequent chapters. It can safely be skipped if desired. The following facet defines a simple architecture that connects two models in sequence:

```
facet sequential[Ti,To,Ty::type]
                (f1,f2::design state_based,
                 x::input Ti,
                 z::output To)::state_based is
  y :: Ty;
begin
  c1: f1(x,y);
  c2: f2(y,z);
end facet sequential;
```

The sequential facet uses a collection of higher-order facet features to define a high-level architecture along with type constraints. Specifically, this is the first facet presented in detail that uses (i) universally quantified parameters, (ii) facets as parameters, and (iii) type parameters.

Most of the interesting work goes on in the facet signature. At first examination, sequential seems to have two parameter lists that differ only in their delimiters — the first using square brackets and the second using parentheses. The first defines universally quantified type parameters while the second defines traditional parameters. The first two parameters in the parameter list, f1 and f2, are defined to be of type state_based, which is also the name of a Rosetta domain. The domain of a facet is also known as its type. Thus, the parameter declaration defines two parameters that must be facets whose domain is state_based.

The facet then uses structural techniques to define a configuration of the two facets passed as parameters. Specifically, f1 consumes system inputs and produces output values that are in turn consumed by f2 and transformed into system outputs. The terms c1 and c2 structurally define the system using the two input facets. Component c1 is instantiated with the system input and component c2 is instantiated with the system output. They communicate via an internal variable, y.

What makes this definition interesting are the types associated with component inputs and outputs as well as the item y that allows components to communicate. When defining a general-purpose architecture, input and output types should not be overspecified. Before facets are assigned to f1 and f2, the types of inputs, outputs, and interconnections are not known. Because the components can exchange literally any item type, our first tendency might be to make their types universal, the supertype of all Rosetta items. However, using the type **top** provides no type information. Once instantiated, we will not be able to check type properties or safely assert anything about the result.

Universally quantified parameters provide a mechanism for achieving the desired result. These parameters serve as place holders for information that will be specified on inferred at a later time when the facet is instantiated. In sequential, three universally quantified parameters are defined to represent the system input type and output type, and the information exchanged between components. None of these values is known when defining the architecture. When the architecture is instantiated with a facet, such information will be available. For the specification to be consistent, values must be found, for each universally quantified parameter, that satisfy type correctness conditions. Consider the following specification fragments:

```
facet absolute(x::input integer; y::output natural)::state_based is
begin
  y' = abs(x);
end facet absolute;

facet square_root(x::input integer; y::output integer)::state_based is
begin
  y' = sqrt(x);
end facet square_root;
posroot(x::input integer; y::output integer) :: state_based is
  sequential(absolute,square_root,x,y);
```

The facets `absolute` and `square_root` calculate the absolute value and square roots of their arguments, respectively. The facet `posroot` uses the sequential architecture to interconnect the two specifications to define a new facet that finds the square root of its input after assuring that the value is positive. Note that in the definition of posroot, only the traditional parameters are instantiated and then only the facet values are instantiated. Universally quantified parameters cannot be instantiated directly — the type-checking system must determine their values automatically.

Knowing the values for `f1` and `f2`, we can now infer values for `Ti`, `To`, and `Ty`. `Ti` and `To` are easy because they are constrained in only one place. `Ti` is the type associated with the sequential facet input used to instantiate the first parameter of `f1`. Because the type of `fi`'s first parameter is `integer`, `Ti` is also integer. The same logic applies when determining `To` is also `integer`.

The interesting case is determining a value for `Ty`, the type of the interconnection between the facets. Both `absolute`'s second parameter and `square_root`'s first parameter provide type constraints on this value. Specifically, `y` must be both a `natural` and an `integer`. A unifying type must be found for `Ty` that accounts for all outputs from `absolute` and can be handled by `square_root`. In this case, `integer` is the desired type and inferring the type of `Ty` is trivial.

The process described informally here is known as *unification*. If no satisfying value can be found for a universally quantified parameter following instantiation, then an error results. Such a case occurs if the first component in the example were to output a character value rather than a natural value. In this case, no type exists for `Ty` that is compatible with both `character` and `integer` other than **top**. The ability to discover such inconsistencies is vital to system level design, where many parameters may not be known during early, abstract design stages.

Defining Facets

Facet definitions nearly always take one of two forms: (i) item declarations or (ii) direct definitions. Item declarations use the same form as used with other Rosetta items. Specifically, a label, type, and value are used to create an item declaration of the form:

$$f :: T \text{ is } v$$

where f names the new facet, T is the facet's type, and v is the facet's value. This style is used frequently when the value of a facet is calculated from other facets using the facet algebra or when defining a facet variable whose value is unknown.

Like functions, this definition style can be cumbersome and difficult to read in many circumstances. Thus, a special syntax is provided to define facets directly. Called *direct definition*, this syntax defines a template over facet declaration elements that results in the common facet definition style seen in previous examples. This template takes the following general form:

```
facet f(parameters) :: T is
  decls
begin
  terms
end facet f;
```

Using the direct definition style makes defining components simple and readable. The following definition for inc uses this style to define a simple state_based model that increments its input and outputs the result:

```
facet inc(x::input integer; z::output integer) :: state_based is
begin
  update: z' = x+1;
end facet inc;
```

10.1 Direct Facet Definition

In Chapter 9 we saw several examples of prototypical facet definitions. Each of these definitions uses the most common mechanism for defining facets, the special syntax for direct definition. Like function direct definition, facet direct definition creates a new item and assigns a value to it in one syntactic expression. Here, the item being defined is a facet and the value is a facet value.

The general syntax for a direct facet definition is:

```
⟦ use  P  ;  ⟧*
facet  F  ⟦ [ variables ] ⟧⟦ ( parameters ) ⟧ :: domain is
  ⟦ ⟦ export ( all | exports ) ⟧
  declarations ⟧
begin
  ⟦ terms ⟧
end facet  F ;
```

where *F* is the facet name, *variables* is an optional list of universally quantified parameters, *parameters* is a list of parameter declarations, *declarations* is an optional set of declarations preceded by an optional *export* clause defining visibility outside the facet, *domain* is the modeling domain, and *terms* is an optional collection of terms. The facet definition is opened by the **facet** keyword and terminated by **end facet** and *F*, where name must be the same in both locations. The facet definition may be preceded by a collection of **use** clauses that specify packages to be included in the definition. The scope of all declarations, parameters, and imported packages is the region between the facet and end keywords.

10.1.1 Parameters

Facet parameter declarations have the following form:

names :: ⟦ *kind* ⟧ *T*

where *names* is a comma-separated list of one or more parameter names, *T* is the parameter type, and the optional *kind* describes a constraint on the parameter. A parameter list is a semicolon-separated list of declarations. The **is** clause is not allowed in parameter declarations.

The *kind* qualifier has three built-in values, **input**, **output**, and **design**. The **input** and **output** kinds are used to label system inputs and outputs, respectively. The **design** kind identifies a parameter much like a generic parameter in traditional design languages. The distinction is that input and output parameters are observed and driven by the system during its operation. Design parameters are system settings that do not vary during operation. The specific semantics of each kind varies with the domain in question.

It is possible to define a parameter without a kind. Such parameters are unconstrained and may be used as inputs, outputs, design parameters, or any other system parameter. Such parameters are useful when direction is not meaningful in describing a parameter or the parameter may represent bi-directional flow. Designers are encouraged to use kinds whenever possible to simplify both specification and analysis.

10.1.2 Universally Quantified Parameters

A universally quantified facet parameter declaration has the same syntax as a parameter has, except no kind can be specified. Parameters and items defined as universally quantified parameters differ in that the latter need not be instantiated when the facet is used. Instead, their values can be determined using type inference and unification techniques during facet instantiation and type checking. Consider the following definition of a component that adds two values and outputs the result:

```
facet adder
  [a::subtype(number)]
  (x,y::input a; z::output a)::state_based is
begin
  z' = x+y;
end facet adder;
```

In this definition, the type of the input and output parameters is the parameter, a, defined to be a subtype of number. When the facet is instantiated in another facet definition, the type of a must be determined. For example:

```
facet intAdder(x,y::integer; z::integer)::state_based is
begin
  adderc : adder(x,y,z);
end facet intAdder;
```

In this declaration, type checking would determine that the value of a must be integer through the unification process. Because integer is in fact a subtype of number, it is a legal instantiation of a. Here, the type associated with a assures that an addition operator exists for the actual value.

Although the following facet looks legal, we cannot establish its correctness:

```
facet wordAdder(x,y::word(8); z::word(8))::state_based is
begin
  adderc : adder(x,y,z);
end facet wordAdder;
```

The inference process can determine that the value of a must be word(8). However, word(8) is not a subtype of number and cannot be a value of a. In this situation, the adder can be instantiated only with number types.

When directly instantiated, universally quantified parameters behave like traditional facet parameters. If a specification frequently instantiates a universally quantified parameter, making it a traditional parameter should be considered. A universally quantified variable is needed only when the specifier needs the type inference system to determine a type automatically.

10.1.3 **Declarations**

Local items are defined within the declaration section along with an optional
export clause. Local item declarations follow the format described in Chapter 4.
A list of item names is followed by a declaration:

names :: T ⟦ **is** *expression* ⟧ ;

Any item type may be defined within the declarative section, including variables
and constants, functions, types, packages, and other facets.

Immediately following the domain and preceding the collection of declarations
is an optional **export** clause having the following format:

export ⟦ *names* | **all** ⟧

The **export** clause defines what internal facet definitions can be seen from
outside the facet. The **export** keyword is followed either by a list of item names or
the keyword **all**. If the **export** clause appears with the **all** keyword, then all declarations defined in the facet are visible. This includes declared items and labeled
terms. If the **export** keyword appears with an export list, then only items listed in
names are visible. Any item, parameter, or term defined in the specification, not
just those in the declarative region, is eligible for export, including term names.
If the **export** clause is omitted, then no items defined within the facet are visible
outside the definition.

The export definition is strict, with no mechanism for overriding the **export**
clause when the facet is instantiated. Specifically, if an item is not exported from
a facet, there is no mechanism for referencing the item in the including scope. If
a need exists to see an item declared in a facet, then it must be explicitly exported
or implicitly exported using the **all** keyword.

To access definitions within a facet, the common "dot" notation is used. Given
a facet name and an item name within the facet, the notation:

facet.item

is used to reference the internal item. Similarly:

$facet_0.facet_1.facet_2.item$

refers to an item declared three levels deep within a definition. It is important
to note that at each level, the referenced *item* must be exported. If *item* is not
exported from $facet_2$, then the reference chain breaks down, as the item cannot
be referenced.

10.1.4 **Domain**

The required *domain* identifier is used to identify the facet's underlying semantics
by specifying *units of semantics*, *model of computation*, and *vocabulary*. The units

of semantics define the basic constructs over which computation is defined. Two units of semantics common in Rosetta specifications are *state based* and *signal based*. The state-based unit of semantics specifies that each facet using it must define the concepts of state and next state. In contrast, the signal-based unit of semantics requires the specifier to define events and associated values. One way of thinking about the units of semantics is as a domain of discourse for specifications using the domain.

The model of computation uses the units of semantics to define how computation is performed. Domains such as `finite_state` and `continuous_time` specialize the `state_based` units of semantics to define finite, discrete state computation, and infinite, continuous computation. Both domains achieve this by assigning specific properties to the state type and next state functions provided by the `state_based` unit of semantics domain.

The vocabulary provided by a domain provides definitions for **common** quantities used in the domain. For example, the `state_based` domain provides a definition for `event(x)` that specifies when an item changes value moving from one state to the next. The `continuous_time` domain defines an item, t, that indicates the current time. Where the units of semantics and model of computation remain anonymous in most specifications, specifiers use definitions provided in the vocabulary extensively.

The domain associated with a facet is also referred to as the facet's type. Thus, the declaration:

```
f :: state_based;
```

defines a new facet using the `state_based` domain. Like any other item declaration, we will read this as "f of type state based." The type associated with a domain is defined as all possible facets written by extending the domain. Thus, any two facets that use the same domain are of the same type. When one domain extends another, that domain is a subtype of the original domain. Because `finite_state` extends `state_based`, `finite_state` is a subtype of `state_based`.

The domains defined in the extended Rosetta domain set are shown in Figure 10.1. Domains are shown in a semi-lattice structure that is called the *domain semi-lattice*. The semi-lattice is arranged hierarchically, with arrows representing extensions. For example, the `discrete` domain is defined by adding definitions, or extending, the `state_based` domain. Extensions define subtypes, thus `discrete` is a subtype of `state_based` and `state_based` is a supertype of `discrete`. Relationships are transitive, thus `finite_state` is also a subtype of `state_based`. Domains, their uses, and semantics are discussed extensively in Chapter 12.

10.1.5 Terms

Terms specify facet behavior by either defining a facet's properties or constituent components. When defining properties directly, terms describe everything from

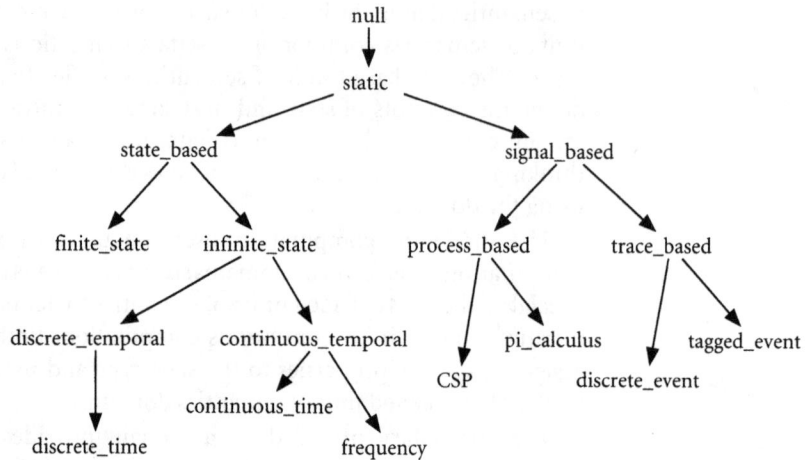

Figure 10.1 Domains defined for the base Rosetta language.

high-level requirements and constraints to fully executable descriptions. When defining components, terms instantiate, rename, and interconnect other facets to form structural definitions. Because terms are declarative, they are not ordered and define a set of properties. Also, terms are not necessarily executable; however, in some restricted cases terms are executable. In the most general case, terms simply define true properties and components.

A facet may contain any number of terms having the following form:

〚 *l* : 〛 *e* ;

where *l* is the term's name and *e* is either a boolean valued expression or a facet valued expression. If the term is a boolean property, the label may be omitted. If the term is a facet, the label must be present. Like all Rosetta declarations, term declarations define items. In this case, the item type is either boolean or facet, the item value is defined by *e*, and the term can be referenced by *l*.

For example, the following term states that inc(3) is equal to 4:

```
l1: inc(3) == 4;
```

The term label is l1 and the expression is inc(3)==4. The semicolon terminates the term definition; however, it does not indicate any type of sequencing of terms. Term order in a facet is immaterial.

The label and semicolon define the term's scope. The label opens the scope while the semicolon terminates the labeled expression. Thus, the specification fragment:

```
l1: inc(3) == 4;
l2: 1 in {1,2,3};
```

defines two terms with labels l1 and l2 and term expressions inc(3)==4 and 1 in {1,2,3}, respectively. It is equivalent to:

```
l2: 1 in {1,2,3};
l1: inc(3) == 4;
```

because terms are simply boolean declarations. The specification fragment:

```
l1: inc(3) == 3;
l2: 0 in {1,2,3};
```

is also a statically legal term list even though it is inconsistent. In contrast:

```
l1: inc(3) == 4
l2: 1 in {1,2,3};
```

is not legal because the first term is missing a terminating semicolon.

Semantically, the semicolon behaves as a conjunction. Terms delineated by semicolons in the body of a specification are simultaneously true and form the set of terms associated with the facet. A facet is consistent if and only if its domain, terms, and declarations, including type conditions, are mutually consistent.

Facet definitions may seem quite similar, but with proper interpretation mean quite different things. The following examples demonstrate this fact by showing how similarly defined terms have different semantics based on the definition domain used by the excluding face.

The following term l1 asserts that x is equal to f(x):

```
facet inconsistent::static is
  x::integer;
  f(i::integer)::integer = i+1;
begin
  l1: x = f(x);
end facet inconsistent;
```

The domain for this term is static, referring to Rosetta's basic mathematical system. There is no concept of state, time, or change in this domain. Thus, x=f(x) is an assertion about x that must always hold. This domain is frequently termed the *monotonic* domain because change is not defined; f(x) is never equal to x because no integer is equal its successor. Thus, this term is inconsistent and the specification is in error.

In the following facet, the terms and declarations remain the same. However, the domain is changed to state_based. This domain allows for values to change by defining concepts of state and next state:

```
facet inconsistent::state_based is
  x::integer;
  f(i::integer)::integer = i+1;
begin
  l1: x = f(x);
end facet inconsistent;
```

Unfortunately, the inconsistency remains because x refers specifically to x in the current state. 11 asserts that x in the current state is equal to f(x) in the current state. The specification is semantically no different than the first.

The following facet fixes the problem by taking advantage of the state_based domain:

```
facet consistent::state_based is
  x::integer;
  f(i::integer)::integer = i+1;
begin
  l1: x' = f(x);
end facet consistent;
```

The notation x' is defined by the state_based domain to refer to x in the next state. Term 11 now asserts that x in the next state is equal to x in the current state plus 1. This is the semantics that x=f(x) has in a traditional imperative language and the semantics we want here.

With some experience, inconsistencies such as those just described are relatively easy to find by hand or with tool assistance. Finding inconsistencies automatically is an exceptionally hard problem with no known solutions. Thus, one must explore specifiations by hand or by guiding tools manually.

The following facet asserts that x at current time plus 5ms is equal to f(x) in the current state:

```
facet sample::continuous_time is
  x::integer;
  f(i::integer)::integer = i+1;
begin
  l1: x@(t+5e-3) = f(x);
end facet sample;
```

This specification is quite similar to the previous specification in that the value of x in some future state is equal to f(x). It differs in that the specific state is referenced using the @ operator and it uses the continuous_time domain. The notation x@t refers to the value of x in some state t. Specifically, the term 11 asserts that in the state associated with 5ms in the future, x will have the value associated with f(x) where the argument to f is the value of x in the current state.

The definition of x' from facet consistent is a shorthand for the definition x@next(s). The relationship between x and x' used across computer science and engineering is that x refers to the current state while x' refers to the next state. The definition of x'using next provides a precise semantics for this common shorthand.

Using the continuous_time domain introduces a different semantics for state. The state_based domain in the previous facet defines state and change of state, but does not assign specific semantics to state. The continuous_time domain goes farther, specifying that state is observed as a continuous value associated with time.

EXAMPLE 10.1

Constant Definition
Shorthand

In addition to demonstrating what a domain is, these examples illustrate a fundamental feature of the Rosetta language. Where traditional programming and design languages embed the semantics of the computation model in the language definition, Rosetta exposes it to the user. Furthermore, the specifier may use different underlying computation semantics when developing system models. This capability is a fundamental step toward true system-level modeling.

The standard definition syntax for constant items is actually a shorthand whose meaning is defined in terms of function declarations and terms. For example, the following definition can be used to define a constant value for pi:

```
pi :: real is 3.1416;
```

The definition is a shorthand form of:

```
facet pi-definition::static is
  export all;
  pi::real;
begin
  pi_def: pi = 3.1416;
end facet pi-definition;
```

Constant functions are defined similarly. The definition of inc is:

```
inc(x::integer)::integer is x+1;
```

An alternative, semantically equivalent definition assigns a specific function to a function variable:

```
facet inc-definition::static is
  export all;
  inc(x::integer)::integer;
begin
  incdef:  inc =} <* (x::integer)::integer is x + 1 *>;
end facet inc-definition;
```

This definition is identical to the standard definition above. However, the shorthand definition of constants and constant functions is more readable and easier for machines to recognize and process. Furthermore, it extends the definition conservatively, assuring consistency of any resulting definition. Whenever possible, the shorthand notation should be used for defining constant values and functions.

The universal quantifier can also be used to provide function definitions. The following example defines a function variable and uses universal quantification to define the function's behavior:

```
facet comprehension::static is
  export inc;
  inc(x::integer)::integer;
begin
  incdef: forall(x::integer | inc(x) = x + 1);
end facet comprehension;
```

The definition states that for every x taken from `integer`, `inc(x)=x+1`. This is semantically equivalent to previous definitions and provides significant flexibility. However, it is exceptionally difficult for machine interpreters to determine when functions defined this way are constant. ∎

In addition to the **let** expression, the facet term definition language provides a **let** form for sharing declarations across multiple terms. Consider the following definition:

```
let x::integer be g(a) in
  l1: f(x,5);
  l2: g(x)
end let;
```

Although syntactically similar to the **let** expression, the term **let** is semantically quite different. It behaves more like a statement than an expression. When the previous **let** term is evaluated, the following terms result:

```
l1: f(g(a),5);
l2: g(g(a));
```

The item declared in the **let** term is simply replaced by its value in the enclosed terms. Like the **let** expression, multiple variable declarations and nested **let** forms are allowed.

The **let** term used in the terms section defines items local to a collection of terms not visible outside the **let**'s scope. In the previous example, the item x defined in the **let** term cannot be referenced in any terms other than l1 and l2. Furthermore, the required presence of the **be** clause implies that x is a constant and cannot be used to define communication between components. For example, the following definition of a 2-bit adder is not legal:

```
facet adder2_fcn(x0,x1,y0,y1::input bit, co::output bit,
             z0,z1::output bit)::continuous_time is
begin
  let ci::bit in
    b0: adder_fcn(x0,y0,0,z0,ci);
    b1: adder_fcn(x1,y1,ci,z1,co);
  end let;
end facet adder2_fcn
```

The local definition, ci, is defined in the **let** term, but no **be** clause is provided to define a value. If this problem is corrected by providing a value, c1 cannot be used to communicate because its value will be constant.

Formally, the **let** term has the following syntax when used to define local variables for terms:

```
let name0 :: T0 be e0
    [ ; namek :: Tk be ek ]* in
  [ term ]*
end let;
```

Note that unlike the **let** expression, multiple names in a single declaration are not allowed. The semantics of **be** is identical to **is** and the **be** clause must be included in each declaration.

10.2 Separable Definitions

Rosetta provides a mechanism for separating the definition of any facet into its *interface* and *body*. The interface defines parameters, a domain, and potentially exported declarations. In addition, use clauses associated with the interface define required packages for the specification. The body defines terms and declarations that will not be exported. No new facet parameters, domains, declarations, or use clauses may be defined with in a package body.

As an example, the following definition for a half-adder can be split into an interface and body:

```
facet halfAdder
  (x,y::input bit; z,cout::output bit)::state_based is
begin
  sum: z' = x or y;
  carry: cout' = x and y;
end facet halfAdder;
```

The interface of this definition allows the user to see a black box representation of the device. Enough information must be provided to allow a user to include the device in a design without reference to the body. The interface definition provides usage information without reference to the complete specification:

```
facet interface halfAdder
  (x,y::input bit; z,cout::output bit)::state_based is
end facet interface halfAdder;
```

The domain and parameters are declared, allowing the user to connect the adder to other facet definitions.

The facet body provides terms to complete the facet definition. Only terms are added by the body, making it dependent on the interface declaration for parameter and domain declarations:

```
facet halfAdder body is
begin
  sum: z' = x or y;
  carry: cout' = x and y;
end facet body halfAdder;
```

Together, the half-adder interface and body provide the same component as provided by the original facet. The distinction is that specifics of the definition are isolated from the user. Furthermore, the half-adder interface can be defined well before the body, allowing its inclusion in designs without requiring detailed specification.

The syntax of a facet interface declares the facet by specifying its signature and associated use clauses. It borrows heavily from the traditional facet declaration syntax:

```
[ use  P  ; ]*
facet interface  F  [ [ variables ] ][ ( parameters ) ]  ::  domain  is
   [ [ export ( all | exports ) ]
      declarations ]
end facet interface  F  ;
```

The only distinction between the interface definition and a full facet definition is the exclusion of the **begin** keyword and associated terms. The **interface** keyword is added to the declaration to assert that only the interface is being defined. Any item visible in the interface, including those resulting from using packages and including domains, is automatically visible in the facet body. There is no need to re-use packages or domains, or redeclare parameters or local definitions.

The syntax of a facet body simply adds terms and other local definitions to the interface:

```
[ use  P  ; ]*
facet body  F  is
   [ declarations ]
begin
   [ terms ]
end facet body  F  ;
```

Definitions of parameters, domains, and exported declarations are excluded from the body definition. However, use clauses can be included to use packages not specified by the interface. Unlike use clauses associated with the interface, use clauses associated with the body do not extend over the interface. This allows the body to locally specify packages that need not be visible to the facet user.

A facet interface does not require the presence of a facet body. If no body is visible when the facet is used, only information from the interface is available for analysis. Effectively, it is like defining a facet with no terms. Because Rosetta is not executable, this situation is perfectly acceptable. Many kinds of static analysis can be performed knowing only information available in the interface.

The declaration of a facet body requires the presence of an associated interface. Because the interface defines items and includes packages used in the body definition, the interface must be present. Only one body can be defined in the scope of any interface. Rosetta has no mechanism for distinguishing between body definitions.

EXAMPLE 10.2

Using Facet Interfaces to Define a Black Box view of a TDMA Signal Processing Component

Among the most common tasks in systems design is designing block-diagram-level specifications. Virtually every systems design begins with a block diagram describing interactions such as information flow between system components. Using facet interfaces is an excellent way to define such diagrams formally without including too much detail.

```
facet interface resampler
  (i::input complex; o::output complex; clk::input bit)::discrete_time is
end facet interface resampler;

facet interface carrierRecovery
  (i::input complex; o::output complex; clk::input bit)::discrete_time is
end facet interface carrierRecovery;

facet interface decimator
  (i::input complex; o::output complex; clk::input bit)::discrete_time is
end facet interface decimator;

facet interface bitSynchronization
  (i::input complex; o::output complex; clk::input bit)::discrete_time is
facet interface bitSynchronization;

facet interface errorCorrection
  (i::input complex; o::output complex; clk::input bit)::discrete_time is
end facet interface errorCorrection;

facet interface messageProcessor
  (i::input complex; o::output complex; clk::input bit)::discrete_time is
end facet interface messageProcessor;
```

The first collection of specifications defines the interfaces for each processor component. The types of the interface ports are defined, and the facets are named and given domains. However, specification details are omitted.

```
facet TDMAstruct
  (i::input complex; o::output complex; clk::input bit)::discrete_time is
    export power;
    power :: real;
    ec2mp,bs2ec,d2bs,cr2d,r2cr :: complex;
begin
  c1: resampler(i,r2cr,clk);
  c2: carrierRecovery(r2cr,cr2d,clk);
  c3: decimator(cr2d,d2bs,clk);
  c4: bitSynchronization(d2bs,bs2ec,clk);
  c5: errorCorrection(bs2ec,ec2mp,clk);
  c6: messageProcessor(ec2mp,o,clk);
end facet TDMAstruct;
```

The TDMAstruct facet defines interconnections between the previously defined facet interface definitions. Information flow between components is defined without committing to particular specifications. ∎

Details for elements of the structural TDMA definition can be added by associating a body with any of the facets comprising the specification. The following example is a trivial Rosetta facet body describing the decimator component:

EXAMPLE 10.3

Using a Facet Body to Define a Specification for a TDMA Signal Processing Component

```
facet body decimator is
begin
  o' = if event(clk)
          then decimate(i)
          else o
        end if;
end facet body decimator;
```

The `decimator` body is immediately associated with the `decimator` interface defined previously. The specification body can be changed and further refined without impact to the interface specification. As the design is refined, the specification can be analyzed to determine if refinements violate system-level specifications. ∎

10.3 Facets and Hardware Description Languages

Rosetta terms are unordered and declarative, meaning that there is no notion of execution or execution order. Although this concept is strange to many designers, its embodiment in traditional hardware description languages is quite comfortable. Consider the following signal assignments from VHDL:

```
s <= s+1 after 5ns;
r <= r+1 after 4ns;
```

Although these terms are evaluated in order, they affect their associated signals in the order specified by their associated timing delays. The value of s in the simulation is updated after the value of r as specified by their associated delays. Interestingly, reversing the evaluation order has no affect on the signals. Specifically:

```
r <= r+1 after 4ns;
s <= s+1 after 5ns;
```

results in the same effect on the modified signals.

The equivalent Rosetta specification has the following form:

```
s_update: s@t+5e-6 = s+1;
r_update: r@t+4e-6 = r+1;
```

The distinction between these definitions is in the mechanism used to specify the next state. In the VHDL model, a single time model is used implicitly. In Rosetta, the timing model can be exposed and utilized. The code fragment above simply makes the time value and its manipulation explicit. Like VHDL, this term pair has the same effect regardless of ordering because of the explicit definition of the next state.

Facet terms can be used to define structural representation of components by including and instantiating other facets. Assuming that `adder1` defines a full adder, terms may be used structurally to define a 2-bit adder as follows:

```
bit0: adder1(x0,y0,0,z0,c);
bit1: adder1(x1,y1,c,z1,co);
```

In this definition, `bit0` and `bit1` are names for two `adder1` facets included in the facet definition. In this case, facet parameters are instantiated to provide external communication through external ports `x0` and `z0` or communication between devices within the component through shared parameters such as `c`. The 2-bit adder will be explored extensively in the discussion of structural definition that follows. For this discussion, it is enough to understand the basic concept of structural definition.

Unlike traditional hardware specification languages, single models may exhibit both structural and behavioral characteristics. It is possible to mix terms defined by boolean expressions and terms defined by facet expressions. Remember that both sets of terms must be simultaneously true. This capability unique to Rosetta allows users to define both the structure and the properties of a system in the same facet. It will become clear shortly that although powerful, this technique is not always appropriate. This is particularly true when properties are defined in distinct domains such as design constraints and functional definitions.

10.4 Facet Styles

When a facet contains only boolean-valued terms, the facet is defined by defining properties directly. Such definitions are referred to as *property-based* specifications and correspond to behavioral modeling in traditional specification languages. When a facet contains only facet-valued terms, the facet is defining *components* and *interconnections*. Such definitions are referred to as *structural* specifications and correspond to architecture or structural specification in traditional specification languages. When the facet contains both boolean-valued and facet-valued terms, the definition is referred to as a *mixed* definition, where both structural and property based definitions apply. Because Rosetta is not interpreted, asserting properties mixed with structural specifications does not present the same kinds of issues as presented by operational specification languages.

10.4.1 Property-Based Facets

When facets contain terms that are labeled boolean expressions, the terms are simply assertions of properties over items. For a facet to be consistent, the terms defined in the facet and the terms defined in its domain must be mutually consistent. Specifically, `false` should never be derivable from a term collection.

Consider the term defining a relational property over an item:

```
t1: x =< 5.0;
```

This definition simply asserts that the value of `x` is less than or equal to `5.0`. It is not conditional nor does it represent an assignment or executable statement. It simply states that the value of `x` is less than or equal to `5.0` in the context of the definition. In the facet definition:

```
facet f(x::input real, z::output boolean,
        d::real design)::state_based is
begin
  t1: x =< 5.0;
end facet f;
```

the term states that the input parameter x must be less than or equal to 5.0, effectively defining a usage assumption over x. If an input greater than 5.0 appears on x, this term cannot be true. Thus, the composition of this facet with any facet that causes a value greater than 5 to appear on input x is inconsistent. Even when both original facets are consistent, the composition may be inconsistent.

A similar term uses equality to constrain the value of x in a different manner. The term:

```
t2: x = 5.0;
```

states that x is equal to the value 5.0. This is not an assignment statement, but simply asserts that x is mathematically equal to 5.0. Anywhere in the context of this definition, x can be replaced by the value 5.0. Using this definition, the following pair of terms is inconsistent:

```
t2: x = 5.0;
bad: x = 6.0;
```

To determine why this pair of terms is inconsistent simply recall that (i) terms are unordered in a facet and (ii) equals asserts equality and is not an assignment. Borrowing from traditional programming languages, one might assert that these two statements assign 5.0 and 6.0 to x, resulting in a final value of 6.0. However, these two terms state that x is equal to 5.0 and 6.0, respectively. Simple substitution then generates the equality 5.0=6.0, which is known to be false. Thus, the term pair is inconsistent and cannot be used in a facet definition.

A similar problem is introduced with the term:

```
bad: x = x+1;
```

This common programming structure assigns the value x+1 to the variable x following its execution. However, in Rosetta, equals is not assignment but an assertion that two values are the same. In this case the term states that the value of x is equivalent to x+1, a provably false statement. Unlike traditional programming and hardware description languages, Rosetta forces the designer to indicate when an equivalence must hold. The previous bad term interpreted in languages such as C implicitly indicates that x is incremented *in the next state*. Rosetta simply requires that specifiers explicitly indicate when symbols are interpreted as being in the current or next state. The "tick" decoration used earlier is a shorthand notation for "in the next state." Thus, the new and correct term can be expressed:

```
good: x'= x+1;
```

which is interpreted as "x in the next state is equal to x in the current state plus 1." Using this simple notation, Rosetta terms can be used to define relationships between the current and next state within a facet. This extremely powerful capability allows consistent and convenient definitions of properties within a facet. Whether working in discrete, continuous, or spatial domains, the concepts of the current and next state are powerful and meaningful in the specification.

10.4.2 Structural Facets

Structural definition in Rosetta occurs when all terms are facets. In the following definition, two adder components are composed within a facet to create a 2-bit adder:

```
facet adder2(x0,x1,y0,y1,ci::input bit,
             z0,z1,co::output bit)::static is
  c::bit;
begin
  b0: adder1(x0,y0,ci,z0,c);
  b1: adder1(x1,y1,c,z1,co);
end facet adder2;
```

In this definition, b0 and b1 relabel adder1 facets that perform addition on two bits. The first adder's data inputs are instantiated with facet parameters x0 and y0 while its output is instantiated with facet parameter z0. Its carry-in value is instantiated with facet parameter ci and its carry out is instantiated with the internal variable c. The effect is defining interconnections between the internal components and the enclosing facet's interface.

The two terms include and rename two copies of the adder1 facet that functionally represents the behavior of a 1-bit adder. The term:

```
b0:adder1(x0,y0,ci,z0,c);
```

creates a copy of the adder1 facet, renames it b0, and instantiates it with system parameters and an internal variable. This process is called *relabeling* and allows the inclusion of the same facet by giving each copy its own name. In this example, two 1-bit adders are included in the definition, but each adder is distinct, as it is a renamed copy of the original definition. Using the previously defined notation, adder2.b1 refers to the adder named b1 inside adder2. Because the 1-bit adder is renamed, the notation adder2.adder1 is not defined.

Figure 10.2 graphically shows the structure of the 2-bit adder defined in the previous example. Parameters from the adder2 facet interface are used to instantiate parameters of the two 1-bit adders. This instantiation causes the 1-bit adders to share symbols with the interface. Thus, when x0 is instantiated when adder2 is used in a definition, the parameter in b0 is instantiated with the same item. Thus, inputs to b0 and b1 change when their associated parameters in the adder2 interface change. Likewise, when the outputs of b0 and b1 change, adder2 outputs instantiated with the same item also change. In this way, facet input and output values are communicated to internal facet interface parameters.

The internal variable, c, is used to facilitate communication of a carry value from b0 to b1. It works in a manner similar to that of interface parameters, in that

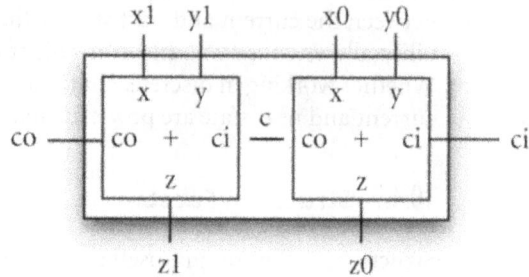

Figure 10.2 A structurally defined 2-bit adder.

when one facet constrains the port instantiated with c, the constraint information is immediately available to the other facet. Specifically, when b0 sets the carry value, b1 sees the result because it shares a the item, c.

10.4.3 Mixed Facets

As the name implies, mixed definitions contain both structural and property-based terms. Where structural and behavioral definition with the same block in VHDL is illegal, Rosetta takes an approach closer to Verilog, allowing structural and behavioral constructs in the same specicification. Combining structural and property-based specification within the same facet in Rosetta is encouraged and useful for defining properties of facet assemblies. Specifically, the properties specified by boolean terms are simply assertions over the same items as used by the structural definition. Thus, the property-based assertions can be thought of as defining conditions that must hold over the assembly of facets. Consider the definition of a power constraint over a component assembly using a mixed definition:

```
facet adder2_const::static is
  export p;
  p :: real;
begin
  b0: adder1_const;
  b1: adder1_const;
  c: p = b0.p+b1.p;
end facet adder2_const;
```

In this facet definition, two instances of the adder1_const facet are included and renamed to represent power constraints defined for the two 1-bit adders. The structural terms b0 and b1 structurally define the power model of the 2-bit adder.

The boolean term, c, defines the power of the structural component as the sum of the component powers.

10.5 Scoping Rules

Rosetta is a statically scoped language, thus the scope of a definition and the item referred to by an item reference can be determined by the syntax of a specification. To understand scoping rules, one must understand the four sources of definition information. The *local scope, domain scope, including scope,* and *use clauses* all provide declarations that can be referenced in a facet, package, or component definition.

Defining the symbols available in a facet's scope begins with the domain. Each facet extends its domain with new declarations and terms. Thus, the declarations and terms defined in a domain are there when the facet definition begins. For example, the state_based domain defines a collection of items that are common to all stateful systems:

```
domain state_based::static is
  state :: type;
  s :: state;
  next(s::state)::state;
begin
  ...
end domain state_based;
```

Any facet defined using the state_based domain starts with definitions of the state type (state), the current state (s), and the next state function (next). They are part of the local scope and can be referenced within any state_based facet. Thus, the following definition represents a correct model:

```
facet incrementer
  (vi::input integer; vo::output integer)::state_based is
begin
  vo@next(s) = vi+1;
end facet incrementer;
```

Although s and next are not defined directly in the facet, the declarations are included from the state_based domain.

The inclusion of domains in this manner provides a primitive kind of inheritance among domains and facets. The state_based domain defined here has an associated domain, in this case the static domain. This implies that any declarations in the static domain are also included in the state_based domain and any facet that uses the state_based domain. The discrete and continuous domains inherit from the state_based domain and thus include definitions of state, s, and next. As the domain collection is formed in this way, the type associated with each domain forms a subtype of its domain. For example, the state_based domain type is a subtype of the static domain because the static domain is its domain.

The next addition to symbols in the facet's scope is the local declarative region. This includes the universally quantified parameter list, parameter list, and local declarations between the **is** and **begin** keywords. Any declarations made in this region are visible throughout the facet definition. A second declaration of `incrementer` uses a local function to implement the increment action:

```
facet incrementer
  (vi::input integer; vo::output integer)::state_based is
  inc(x::integer)::integer is x+1;
begin
  vo@next(s) = inc(vi);
end facet incrementer;
```

Within the facet term, the `inc` function is visible and is used to define the increment property. Also, items from the parameter list declaration are used in the term to reference elements of the facet interface.

A common mistake with writing facets is attempting to redefine items declared in the facet's domain. Items in the domain are treated as part of the facet and any attempt to redeclare the item in the local scope is treated as a second, illegal definition. For example, in the previous facet definition, if s and next are defined locally, an error results because the new declarations are repeated declarations in the same scope:

```
facet incrementer
  (vi::input integer; vo::output integer)::state_based is
  inc(x::integer)::integer is x+1;
  s :: state;
  next(s::state)::state;
begin
  vo@next(s) = inc(vi);
end facet incrementer;
```

The local definitions of s and next conflict with the definitions from the state_based domain, resulting in a redeclaration error. It is still possible to constrain items defined in the domain using terms. The type state and the function next are not interpretable in the current definition. They can be given values using terms; however, this practice is discouraged in favor of using a more concrete domain definition. More will be said about this in Chapter 12.

As seen earlier in this chapter, the **use** clause provides a mechanism for including definitions from packages. When a package is referenced in a **use** clause, all declarations from the package are included in the succeeding declaration, qualified with the package name. The definition of the `increment` function from the previous example can be moved into a package and included using a **use** clause as follows:

```
package inc_package is
  inc(x::integer)::integer is x+1;
end package inc_package;

use inc_package;
facet incrementer
```

```
  (vi::input integer; vo::output integer)::state_based is
begin
  vo@next(s) = inc_package.inc(vi);
end facet incrementer;
```

In this definition, the **use** clause includes the definition of inc in the facet definition. The quantified name, inc_package.inc, specifies the definition of inc from the included package. Because there is only one definition of inc in the facet scope, the qualified name may be omitted and the simple name inc used instead.

If a local definition of inc appears in the facet definition, the local definition hides the package definition unless the qualified name is used. This example defines a local version of increment in addition to the definition from inc_package:

```
use inc_package;
facet incrementer
  (vi::input integer; vo::output integer)::state_based is
  inc(x::integer)::integer is x+2;
begin
  vo@next(s) = inc(vi);
end facet incrementer;
```

Because the term uses an unqualified name, the local definition of inc is used and the facet model increments its input by 2 in each state. The definition from inc_package is still present and may be referenced using its qualified name:

```
use inc_package;
facet incrementer
  (vi::input integer; vo::output integer)::state_based is
  inc(x::integer)::integer is x+2;
begin
  vo@next(s) = inc_package.inc(vi);
end facet incrementer;
```

Note that if two or more used packages provide a declaration for the same item, the qualified name must always be used unless a local definition is present. If a local definition is present, then the unqualified name always refers to it.

An interesting example makes the local definition equivalent to the packaged definition:

```
use inc_package;
facet incrementer
  (vi::input integer; vo::output integer)::state_based is
  inc::<*(x::integer)::integer*> is inc_package.inc;
begin
  vo@next(s) = inc_package(vi);
end facet incrementer;
```

Here the value of the local definition is defined to be the value from the packaged definition. Now the local, unqualified definition refers specifically to the definition from inc_package.

It should be noted that the **use** clause appears before the facet declaration. The reason for this is to make clear that declarations from a used package can be referenced in the parameter list and universally quantified parameter list. This is particularly useful when packages contain types that may be used parameter declarations.

The final source of item declarations in a facet is the scope in which the facet is defined. Declarations in the including scope may be referenced in the internal facet's definitions. For example, if the incrementer facet is defined locally within another facet, definitions from the outer facet are visible:

```
facet outer::state_based is
  inc(x::integer)::integer is x+2;

use inc_package;
facet incrementer
  (vi::input integer; vo::output integer)::state_based is

begin
  vo@next(s) = inc(vi);
end facet incrementer;

begin
  ...
end facet outer;
```

In this case, the definition of inc comes from the outer facet. If a packaged declaration of the same name is available, then the unqualified name refers to the declaration from the outer context. Thus, the **use** clause in the previous definition is not referenced in the facet declaration. If a local definition is included, it overrides the definition from the outer context, much like a packaged definition. However, there is no qualified name for the declaration from the outer context and it cannot be referenced in the scope of the new declaration.

10.6 Basics of Facet Semantics

To this point, little has been said about the semantics underlying facet definitions. The details of that discussion are beyond the scope of this book. However, a brief discussion of how that semantics is defined is included here. The mapping of a facet to its underlying co-algebra is presented followed by a discussion of type semantics. This section may be safely skipped by those simply interested in learning about writing Rosetta specifications.

10.6.1 Facet Semantics

Facet semantics are denoted by mapping terms and item declarations to a *co-algebra*. Co-algebraic semantics is chosen over the more popular algebraic semantics due to the reactive nature of Rosetta facets. Each facet observes its inputs and defines a response to changes in those inputs, defining a kind of stream transformer. Additionally, the abstract nature of the co-algebra's state makes defining many different state observations much easier.

A co-algebra is defined by an abstract state, χ, observations on the abstract state, $\vec{\imath}$, and types associated with each observation, $\vec{\tau}$. The signature of a co-algebra defines a collection of typed observations over an abstract state:

$$\vec{\imath} :: \chi \rightarrow \vec{\tau}$$

where $\vec{\imath}$ defines a vector of observations, χ is the abstract state, and $\vec{\tau}$ is a vector defining observer types.

Denoting a Rosetta facet as a co-algebra maps declared items to observations, item types to observation types, and terms to co-algebra terms defining properties of observations. If the abstract state is viewed as system state, each Rosetta facet defines a different observation of the system.

EXAMPLE 10.4

Denotation of a Simple
Rosetta Facet as a
Co-algebra Structure

Given the following facet defining a simple increment operation:

```
facet increment(o::output integer; clk::input bit)::state_based(integer) is
  inc(x::integer)::integer is x+1;
begin
  update: if rising(clk) then s' = inc(s) else s' = s end if;
  out: o' = s';
end facet increment
```

the corresponding co-algebra signature is:

```
<o,o',clk,inc,s,s',next> ::
    χ -> <integer,integer,bit,integer->integer,state,state,state->state>
```

The co-algebra defines a mapping χ to a vector of values associated with the specified type; o is an integer, inc is a mapping from integer to integer, and so forth. Given the co-algebra signature, terms are defined over signature elements:

```
<o,o',clk,inc,s,s',next> ::
    χ -> <integer,integer,bit,integer->integer,state,state,state->state>
    if rising(clk) then s' = inc(s) else s' = s end if;
    o' = s';
```

All items, including the facet state and next state function, correspond to observations of the abstract co-algebra state. Thus, the state Item, s, in a facet does not correspond to the abstract state, χ, in the co-algebra. It is a simple observation identical in every way to any other observation. Thus, the system state can be consistently observed in different ways by different facets. A digital system can be observed as having both an analog and digital state by defining facets where s is of type real and bitvector, respectively. Because these are simply observations of the state and not the state itself, the observation types are not inconsistent.

An interaction between domains relates observations in different facets. If the state in any facet is simply an observation of system state, it is possible to define relationships between these observations. In Rosetta, such relationships are defined using interactions and they move information from one domain to another. Because state is defined at the domain level, interactions can also be defined at the domain level, supporting general interaction styles.

Many specification languages sharing the style used by Rosetta define their semantics algebraically. The denotation of such languages is quite similar to Rosetta mapping declarations in the language to structures in the semantic system. However, algebraic specifications use concrete type for state representation unless a hidden algebra is defined. When using algebraic semantics, the state has a concrete type, making definition of multiple observation types more difficult. The chosen concrete state type will always be biased to one particular state representation. The co-algebraic semantics only defines observations of state, not a concrete state type, avoiding bias caused by a concrete state type.

10.6.2 Facets and Type Semantics

All facets must well-typed, implying that the facet must be properly structured and each declaration must be well-typed. Examining the facet's structure reveals the kinds of checks performed during static analysis. The counter facet defined in Figure 10.3 exemplifies the kind of static analysis necessary to determine if a facet is well-typed.

All declarations in counter must themselves be well-typed, starting with the internal declaration of inc_mod4 and rising. Each declaration defines any function that can be checked using rules defined in Chapter 6. The expression associated with each function must be well-typed and have the same type as its specified return type.

```
facet counter(reset::input bit;
              value::output word(2);
              clk::input bit):: finite_state(word(2)) is
  inc_mod4(x::word(2))::word(2) is
    case x is
      {b"00"} -> b"01" |
      {b"01"} -> b"10" |
      {b"10"} -> b"11" |
      {b"11"} -> b"00"
    end case;
    rising(s::bit)::boolean is event(s) and s=1;
  begin
    next_state: s' = if reset=1
                       then b"00"
                       else if rising(clk)
                              then inc_mod4(s)
                              else s
                            end if
                     end if;
    output: value = s;
  end facet counter;
```

Figure 10.3 Specification of a 2-bit counter with reset.

Terms are declarations and must also be well-typed. The expressions that define each term must well-typed and must be of type `boolean` or a facet type. In each case, `counter` terms define boolean expressions that are in fact well-typed.

Finally, the domain instance must be well typed and be of some domain type. The conditions for a domain instance being well-typed are the same as for a facet being well-typed. Specifically, the type of each actual parameter must be a subtype of its associated formal parameter. In the counter example, the domain used is `finite_state`, whose only argument is a type, or set of values. Because `word(2)` defines the set of two-element bitvectors, the domain instance is well-typed.

Packages, Libraries, and Components

Rosetta provides a collection of capabilities for grouping declarations into composite structures. A *package* is a parameterized Rosetta construct used to group definitions together into reusable collections. Packages represent containers for various related declarations. A *library* defines a location for packages by associating a universal resource indicator (URI) with a local name. A *component* is a collection of definitions that allow inclusion of usage assumptions and verification conditions along with a specification in a single construct. Components allow information, including design rationale, correctness connections, and assumptions for correct usage, to be encapsulated with a facet in a standard manner.

11.1 Packages

Semantically, a package is basically a facet with no terms and whose items are all exported. Every Rosetta definition that is not itself a package must be defined within a package. Thus, packages are the basic element of specification organization. They are used to represent entire projects, component libraries, abstract data types, and collections of related functions and data types. While a facet represents a model, a package represents a related collection of models and other items.

As an example, the following package represents an abstract data type for binary trees:

```
package tree(T::type)::static is
  TreeType(T::type) :: type is data
    nil :: empty |
    node(lt::TreeType,v::T,rt::TreeType) :: nonempty;
  end data;

  size(t::TreeType)::natural is
    if nil(t) then 0 else 1+size(lt(t))+size(rt(t)) end if;
end package tree;
```

This package defines a new constructed type called `TreeType` and a function `size` that evaluates to the size of a tree. The package is parameterized over a type, T, allowing the tree to contain any Rosetta type. To use this package, a **use** clause would include and instantiate the package as follows:

```
use tree(integer);
```

This **use** clause includes the tree package and instantiates the tree element type with `integer`. Thus, the package where the inclusion occurs now has access to a new tree type that contains integers and whose size can be determined.

The **use** clause must appear immediately before the outermost declaration that includes it. Typically, the **use** clauses associated with a construct appear immediately before its definition. For example:

```
use tree(integer);
facet multiplier
    ...
end facet multiplier;
```

includes package elements in the `multiplier` facet. However, the declarations in tree are not visible beyond the **end facet**. The rationale behind this notation is that frequently data structures and functions provided by a package will be used in a construct's parameter list. This is indicated implicitly by including the **use** clause immediately prior to the construct.

This simple Rosetta specification defines a package containing two basic RTL components. The package is defined in the `static` domain; however, the two facets defined within are both `state_based` facets.

```
package rtl()::static is
  facet mux2
    (i0,i1::input bit; z::output bit;
     c::input bit)::state_based is
    mux1: z' = if c=0 then i0 else i1 end if;
  end facet mux2;

  facet demux2
    (e::input bit; z0,z1::output bit;
     c::input bit)::state_based is
    demux0: z0' = if c=0 then e else 0 end if;
    demux1: z1' = if c=1 then e else 0 end if;
  end facet demux2;
end package rtl;
```

To access the package, the **use** clause specifies the package prior to the construct where it is referenced:

```
use rtl();
facet mux4(i0,i1,i2,i3::input bit; z::output bit;
           c0,c1::input bit)::state_based is
begin
m1: mux2(i0,i1,x1,c0);
m2: mux2(i2,i3,x2,c0);
m3: mux2(x1,x2,c1,z);
end facet mux4;
```

The mux4 specification defines a 4-1 multiplexer structurally from three 2-1 multiplexers defined in the rtl package previously.   ▪

11.1.1 Defining Packages

All package definitions have the following format:

```
[ use  P  ; ]*
package  P  ([ parameters ])  ::  D  is
  [ export ]
  [ declarations ]
end package  P  ;
```

The **package** keyword identifies the declaration of a package, while *P* names the package. The optional *parameters* is a collection of parameter definitions. These definitions differ in that all parameters to a package are considered **design** parameters. Typically, such parameters are used to customize characteristics of items provided by the package. The optional *export* allows visibility control over package declarations. By default, all package declarations are exported (unlike a facet, where no declarations are exported).

Like facets, the domain, *D*, provides a vocabulary and language for package specification. Most packages are defined in the static domain to provide maximum flexibility in their use. Note that other definitions within the package do not need to use the same domain as a basis. The domain of the including construct must be a subtype of the included package. This requirement assures that all items defined in the included package domain have meaning in the including domain. For example, it is not possible to include a package of continuous_time functions in a discrete_time domain.

Packages differ from facets in that they have no terms, and all defined items are exported by default. The package body appears between its signature and associated **end** keyword. The **begin** keyword does not appear because no terms will ever follow it. If an **export** clause does not appear in a package definition, all declarations are exported by default. Including an **export** clause results in only those items specified being exported. Using the **export all** notation is the same as having no export clause.

11.1.2 Separable Definitions

Any package definition may be separated into an interface and a body. Like other structures, the interface specifies the visible portion of the package, while the body adds specification detail. Package interfaces have the following syntax:

```
[ use  P  ; ]*
package interface  P  [ parameters ]  ::  D  [ with body ] is
  [ export ]
  [ declarations ]
end package interface  P  ;
```

Use clauses specified with the package interface extend to the package body when the body is present. The package name, parameter list, domain, and declarations are identical to the unified declaration syntax. Items not appearing in the interface cannot be exported and all items are exported by default.

The special syntactic element, **with body**, appearing before the **is** keyword indicates that the interface being defined requires a body. Because packages only provide declarations to other specification units, they do not have associated terms. Thus, an interface that only contains declarations may represent the entire package. If a body should be associated with the interface, the optional **with body** keyword should be included. If an interface is complete without a body, then the keyword should be omitted.

Package bodies have the following syntax:

```
⟦ use  P ⟧*
package body  P  is
    ⟦ declarations ⟧
end package body  P  ;
```

Like other structure bodies, no new parameters, domain elements, or exported declarations can be added. Any packages used by the body are not visible in the interface. Although an interface can be used without a body, a package body cannot be used without its associated interface.

11.1.3 Using Packages

The format of the **use** clause is:

```
use packages;
```

where *packages* is a comma-separated list of instantiated packages.

All exported symbols from packages in *packages* are visible by default in the scope of the use clause without the use of the *name.label* notation. If an exported symbol conflicts with symbols defined in scope or by other included packages, the dot notation is used to indicate the appropriate item. Specifically, if two used packages provide the same symbol, the dot notation is used to resolve the actual symbol used. If a symbol is defined in the construct using it or its associated domain, then the local definition overrides any package definition. If the package definition is required, then the dot notation may be used to reference the specific definition.

It is also possible to include multiple copies of the same package with different parameterization. In this case, dereferencing a name requires the parameters of the package used to be included. For example:

```
use tree(integer);
use tree(float);
```

requires each reference to any symbol in `tree` to include parameters when dereferencing. Specifically:

    tree(integer).nil

refers to the integer tree's nil value, while:

    tree(float).nil

refers to the float tree's nil value. If package names are sufficient to determine the instance associated with a name, then the parameters are unnecessary when dereferencing.

The following package and facet declarations exemplify some name resolution properties:

EXAMPLE 11.2

Package Inclusion and Results of Name Resolution

```
package example1()::static is
  a,b,c::integer;
end package example1;

package example1()::static is
  export c;
  c,d::integer;
end package example;

use example1();
use example2();
facet exampleImport()::static is
  a :: bit;
begin
  a == 1;  // local a
  example1.a == 1;  // a from example1
  b == 2;  // b from example1
  c == 3;  // cannot resolve
  example1.c == 4;  // c from example1
  example2.c == 5;  // c from example2
  d == 6;  // cannot resolve
  example2.d == 7;  // d from example2
end facet exampleImport;
```

11.1.4 The Working Package

All Rosetta items must be declared in a package. However, the package is not always explicitly referenced in the specification model. When Rosetta constructs are defined without explicitly being included in a package, they are included by default in the `working` package. There is no physical package associated with the `working` package. Instead, the contents of the `working` package are located in the current working directory. There is no need to export anything from the `working` package, as it will always be the outermost defining package.

11.2 Libraries

Libraries define virtual locations for packages and other compilation units. A library declaration defines a library name and may associate it with a physical location using a standard universal resource indicator. The library itself is not a location, but rather is a name or alias for a location that is used to reference a collection of compilation units. By using URIs, library locations may be local to the machine or distributed throughout the network.

11.2.1 Library Definition

Libraries are defined in the same manner as other top-level constructs are. The **library** keyword declares a library name and the optional **is** clause provides a value for the library:

library *name* ⟦ **is** *string* ⟧ **end library** *name*;

The *string* value provides a location for the library contents and takes the form of a standard URI pointing to the base library. Some examples of library declarations include:

```
library design_lib end library design_lib;

library local is
   "file:///usr/local/rosetta/local"
end library local;

library rosetta.lang is
   "http://www.rosetta.com/usr/lib/rosetta/lang"
end library rosetta.lang;
```

The library `design_lib` defines a library that is known to exist, but whose location is not known. The `local` library defines a library on the local machine located in:

```
/usr/local/rosetta/local
```

Finally, `rosetta.lang` defines a remote library located on `www.rosetta.com` in the directory:

```
/usr/lib/rosetta/lang
```

Defining libraries in this manner allows distributed system definitions. This is vital for large systems, where design teams are physically distributed over large distances. Using URIs allows Rosetta to take advantage of standard location definition techniques and access methods.

11.2.2 Referencing Library Elements

Library elements are referenced using the standard dot notation. For example:

use local.data_types;

Uses the package data_types located in the local library. If a package appears in a **use** clause without a library qualifier, the working package is assumed with the current location used as the library. For example:

use data_types;

looks for a package named data_types in the current working directory. This is the equivalent of the definition:

use working.data_types;

Here, working is not a library, but rather is the name of the package containing data_types.

11.2.3 Predefined Libraries

Several predefined libraries must exist for any Rosetta installation (Table 11.1). The rosetta.lang library provides basic language definition packages. The prelude package contains the Rosetta prelude, defining all constructs defined in the static domain. If the null domain is extended with the prelude package, the static domain results. The unicode package defines characters and operators for standard Unicode manipulation. Finally, the domains package contains definitions for the base Rosetta domain set.

The rosetta.lang.reflect library provides reflection operators used to define and manipulate Rosetta language constructs. The lexical and abstract_syntax packages contain definitions for lexical constructs and data structures representing abstract syntax, respectively. The semantics, simplification, and name_expansion

Table 11.1 Built-in Rosetta libraries and associated packages

Library	Package	Contents
rosetta.lang	prelude	Base language constructs
	unicode	Unicode definitions and functions
	domains	Base domain definitions
rosetta.lang.reflect	lexical	Lexical constructs
	abstract_syntax	Abstract syntax constructs
	semantics	Semantic definitions for abstract syntax elements
	simplification	Derived form definitions
	name_expansion	Functions for expanding simple names

packages contain semantic definitions, derived forms, and rules for fully expanding abbreviated names into their full forms.

11.3 Components

A *component* groups a traditional facet declaration with usage assumptions and correctness conditions. Each component specifies *assumptions* defining usage assumptions, *definitions* defining the component and playing the same role as facet terms, and *implications* defining correctness conditions. Each of these elements is specified as a collection of terms identical to a set of facet terms.

To illustrate component definition, the following specification defines a binary search component, augmenting the normal functional specification with usage assumptions and correctness conditions:

```
component bin_search(x::input sequence(integer);
                     k::input integer;
                     z::output boolean) :: state_based is
begin
  assumptions
    a1: ordered(x);
  end assumptions;
  definitions
    d1: z'=binSearchFn(k,x);
  end definitions;
  implications
    i1: z'=k in ~(x);
  end implications;
end bin_search;
```

This component specifies that the output in the next state is equal to using the function binSearchFn to search the input sequence. Defined in the **definitions** section, this requirement is the basic functional specification for the component. At the same time, the component makes the assumption that its input is ordered using a term in the **assumptions** section. This is an appropriate guard condition for using a binary search function that requires its input to be ordered. Finally, the component implies that the output in the next state be equivalent to determining if the key is included in the elements of the input sequence. This term appearing in the **implications** section defines a requirement for the binary search component that should be derivable from the system definition. In this way, the definition serves as a correctness condition. It could also serve as the systems's functional requirement if a more abstract specifiation were desired.

Before this component is used in a specification, implications must be supported by assumptions and definitions to ensure correctness. When this component is used in a system specification, the assumptions must be supported by the usage environment while the definitions and implications define correct behavior. The implications can be treated as definitions when usage assumptions are justified because they depend only on local definitions. In a sense, they annotate the component definition, adding details that could be otherwise derived.

11.3.1 Defining Components

The syntax of component definitions has the following form:

```
component  C  ⟦ parameters ⟧:: D  is
   ⟦ export  exports ⟧
   ⟦ declarations ⟧
begin
  assumptions
     ⟦ assumptions ⟧
  end assumptions;
  definitions
     ⟦ definitions ⟧
  end definitions;
  implications
     ⟦ implications ⟧
  end implications;
end component  C ;
```

where *C* names the component, *parameters* is a list of parameters, *declarations* is a set of declarations, *exports* is an export clause, and *D* is the domain of the component. Each of these declarations is identical to the declarations in a facet and plays the same role in the component definition. Where a component differs from a facet is term definition in the body of the specification. The *assumptions*, *definitions*, and *implications* are all collections of terms exactly like those appearing in traditional facet definitions.

The assumptions section defines a collection of terms that are assumed to be true in the component definition. They are not asserted, but instead define usage conditions for the component. Whenever the defined component is included in a definition, its assumptions must hold true in the including system or correct behavior cannot be guaranteed. Assumptions play a role very similar to that of preconditions in an axiomatic specification and record usage conditions for the component.

The definitions section plays the same role in a component as it does in a facet. Specifically, the definitions section declares a collection of terms that define the facet model. As such, they are true within the component and define the component's behavior when used in other systems.

The implications section defines a collection of correctness conditions for the definitions and assumptions. They define terms that must follow from the union of assumptions and definitions and are treated as definition terms by systems using the component. Implications are used when desirable properties should be defined that do not exhibit basic component definitions. Implications are quite useful when performing verification, because they allow verified terms to be included without requiring re-verification.

The following relationship must hold between assumptions, definitions, and implications for a component to be semantically correct:

assumptions ∧ **definitions** ⇒ **implications**

Implications define properties that must be supported by definitions and assumptions. Although ideally this relationship is formal proof, any form of support is acceptable. This reflects the nature of systems engineering, where many reasons can be used to justify a design decision or conclusion. Certainly formal proof is acceptable, but Rosetta semantics does not demand this.

When usage assumptions, definitions, and implications cannot support a formal proof, their inclusion in specifications adds clarity by recording design intent. Many correctness conclusions are made through empirical analysis that are not captured in formal proof. In many cases, conclusions are supported simply by the experiences of the designer. When a design is revisited, understanding assumptions and how they are supported proves vital to successful re-engineering and component replacement.

A shorthand notation is provided for defining a component around an existing facet. Assuming that the facet binSearch existed prior to writing the bin_search component, it could be reused in the component definition as follows:

```
component bin_search(x::input sequence(integer);
                     k::input integer;
                     z::output boolean) :: state_based is
begin
  assumptions
    a1: ordered(x);
  end assumptions;
  definitions = binSearch(x,k,z);
  end definitions;
  implications
    i1: z'=k in ran(x);
  end implications;
end bin_search;
```

Rather than list terms in the **definitions** section, the binSearch facet is used by associating it with **definitions** using the notation:

```
definitions = binSearch(x,k,z);
end definitions;
```

The binSearch facet is simply copied into the **definitions** section without modification. The **assumptions** and **implications** sections can be treated similarly.

11.3.2 Separable Definitions

Like facets, component interfaces may be specified separately from their associated bodies. The syntax for such specifications is virtually identical to that for facet specifications. Specifically, the interface defines use clauses, parameters, the domain, and exported declarations. The body defines terms and local declarations that cannot be exported. The syntax for component interface specification is:

```
[ use  P  ;  ]*
component interface  C  [ parameters ]::D is
    [ export  exports ]
    [ declarations ]
end component interface  C ;
```

Like the facet interface, the component interface parallels the full component specification, except that term specifications are omitted. Use clauses specified over the interface are similarly specified over the body. Thus, all definitions visible in the interface are visible in the body as well.

The component body specification uses the following syntax:

```
[ use  P  ;  ]*
component body  C  is
    [ declarations ]
begin
  assumptions
    [ assumptions ]
  end assumptions;
  definitions
    [ definitions ]
  end definitions;
  implications
    [ implications ]
  end implications;
end component body  C ;
```

Like the facet body specification, the component body specification defines terms over items declared in the component interface. Use clauses specified over the body only do not extend over the interface. Thus, the user need not know about packages associated with the body definition only.

11.3.3 Accessing Component Elements

There are times when it is desirable to reference a declaration or term in a component that is a part of the assumptions or implications definitions. To support this kind of access, the component definition can be thought of as shorthand for the following facet definition:

```
facet  C  [ parameters ]::D is
    [ declarations ]
    export implications, assumptions,  exports;

    facet assumptions::D is
    begin
      [ assumptions ]
    end assumptions;
```

```
facet implications::D is
begin
  ⟦ implications ⟧
end implications;

begin
  ⟦ definitions ⟧
end facet C ;
```

Using this equivalence, it is possible to access the assumptions and implications associated with a component by using the notations *C*.assumptions and *C*.implications. Because these collections of terms are treated as internal facets, outside references can be treated as facet references. Specifically, their terms, parameters, and declarations can be accessed using meta-functions and they can be manipulated using facet expressions and functions.

11.3.4 Using Components

Components are used exactly like facets in structural designs and in facet composition operations. When a component is instantiated, the instantiation is treated as a facet formed from it's definitions. Implications and assumptions are treated as local declarations, but can be referenced using the dot notation as described previously. The reason for allowing this is to support definitions that may include properties from the component that are not explicitly defined as axioms. Because implications are supported by proof or other evidence, they can be referenced and used in other definitions.

EXAMPLE 11.3

Using the binSearch
Component to Define a
Larger Model

The following Rosetta component uses the binSearch defined earlier and a second component, bubbleSort, to structurally model a simple component that finds an integer in a sequence. The bubbleSort component accepts a sequence f integers, sorts it, and outputs the sorted result. This component makes no assumptions about its input, but asserts that its output will be ordered.

```
component bubbleSort(x::input sequence(integer);
                     y::output sequence(integer));
begin
  assumptions
    true;
  end assumptions;
  definitions
    d1: y'=bubbleSortFn(x);
  end definitions;
  implications
    i1: ordered(y');
  end implications;
end component bubbleSort;
```

When the bubbleSort component is connected in series with a binSearch component, a search component that does not require sorted input results. Further, we can be confident about the connection between components because the implication that the bubbleSort output will be sorted satisfies the assumption that the binSearch input will be sorted.

```
component find(x::input sequence(integer); k::input integer;
               z::output boolean)::state_based is
  a :: sequence(integer);
begin
  assumptions
    true;
  end assumptions;
  definitions
    sort: bubbleSort(x,a);
    search: binSearch(a,k,z);
  end definitions;
  implications
    i1: z' = k in x;
  end implications;
end facet find;
```

The find system can be defined as a facet if the assumptions and implications are left out. However, the use of a component records further information about the resulting system. Whether a component or facet is used is up to the user's discretion and often depends on the situation.   ∎

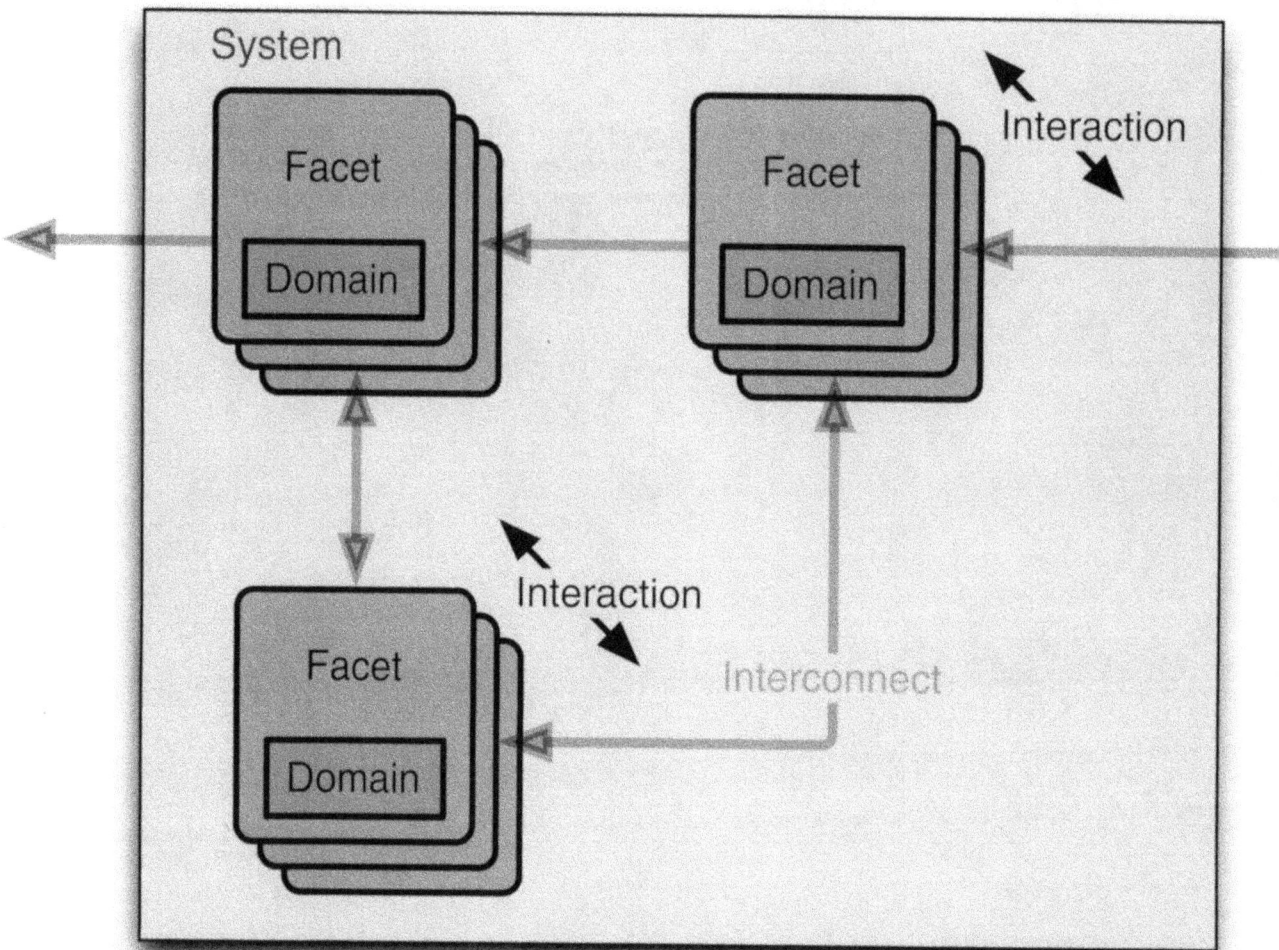

Domains and Interactions

IV

System-level design cannot be performed in a single engineering domain. By definition, system-level design integrates heterogeneous system views to predict effects of local decisions on system-level properties. These local decisions are made by domain specialists using the languages and semantics of their domains.

Part IV describes the Rosetta *domain interaction* systems used to define modeling domains and relationships between modeling domains. Domains describe units-of-semantics, models-of-computation, and engineering design domains by defining vocabulary and semantics used by domain experts. Interactions define how specifications in one domain relate to impact specifications in another. Domains provide the specialist's modeling languages while interactions compose them.

After completing the chapters in Part IV, you will understand how to use domains to define semantic models for facets, how domains are defined, how facets are composed, how functors, translators, and combinators define interactions, and how to use interactions to define system-level impacts of local design decisions.

Domains

The essence of systems engineering is the need to compose and understand specifications from multiple engineering domains. Because different engineering activities use different modeling vocabularies, any language purporting to be a system-level modeling language must support *heterogeneous specification*. More specifically, a system-level modeling language must support and integrate multiple modeling paradigms, models of computation, and communication models.

Rosetta's support for heterogeneous specification is embodied in *domains* and *interactions*. Domains provide a basic modeling vocabulary for a facet and define its type. They allow the designer to choose or define a vocabulary for specification that embodies an appropriate modeling paradigm for that specification. Domains enable Rosetta users to write specifications using different modeling paradigms in the same language system. Interactions describe how heterogeneous specifications interact when combined. Defined over domains, interactions move information from one modeling paradigm to another and are further described in Chapter 15.

We have already seen domains at work in facet definitions, where they are used as common vocabulary for defining specifications. Syntactically, each facet's domain is specified using a type annotation following the facet signature. Packages, components, and other domains use the same definition technique. Semantically, each facet extends its domain by including all definitions from the domain and adding definitions local to the facet. How this extension is realized during elaboration is specified as a part of the domain.

12.1 Elements of a Domain

The primary focus of domain definition is providing a specification vocabulary. Where facets describe specific system models, domains define reusable, common constructs for defining new models. Although domains are like facets in many ways, they are not used as components and cannot be instantiated in structural specifications. They serve as definitional building blocks for facets and group collections of similar facets into types.

Domains accomplish their task by defining *units of semantics, models of computation*, and *engineering domain knowledge* to varying degrees. Units of semantics declare the basic modeling primitives shared by classes of specifications. State-based specifications all reference the current state and the next state function, regardless of their specific properties. Similarly, signal-based specifications all reference signals and processes. Models of computation constrain units of semantics by providing increasingly concrete definitions for them. Finite state specifications constrain the state type to be finite and discrete, but do not define new state vocabulary. Similarly, infinite state specifications constrain the state type to be infinite and say nothing of discreteness. Finally, engineering domains gather information specific to a particular domain, such as digital design, hydraulics, analog design, or power modeling.

Rosetta supports incrementally defining units of semantics, models of computation, and engineering domain information through extension of basic models. Starting with units of semantics, models of computation are defined and extended. Using models of computation, basic engineering domains are defined and extended to sub-domains. As we shall see in Chapter 15, sharing units of semantics and models of computation provides explicit support for specification composition, abstraction, and refinement.

12.1.1 Units of Semantics

The first objective of any domain specification is to provide *units of semantics* declarations. The units of semantics comprise a basic collection of items that represent quantities shared among specifications in a domain. These items take the form of variable, constant, and type definitions. For example, the concepts of state, time, and state change are critical to all state-based specifications. Although these quantities have different properties based on the specifics of a state-based domain, they are always present. Thus, the state_based domain that serves as a basis for all state-based specifications defines these as basic units of semantics. Domains defined from the state_based domain provide different semantics for these items to define different flavors of state_based specifications.

One of the most critical units of semantics definitions specifies the point-of-reference for observation. Examples of such items include state and time as well as position and order. We say that items like state specify the point-of-reference because all observations are performed with respect to them. For example, we specify the values of variables with respect to state in sequential systems or with respect to time in analog systems. Such domain items are not restricted to specifying to time. They may specify spatial position for representing measured values on a surface or events for representing values associated with occurrences.

When a domain provides nothing other than declarations and is defined directly from the static domain, it is informally referred to as a *unit-of-semantics* domain. Most Rosetta users will never define a new unit-of-semantics domain, but will instead extend existing domains to provide new models of computation or engineering domains. The unit-of-semantics domains must be added with great care. As we shall see in Chapter 15, carefully choosing common items among specifications can profoundly influence the utility of a system specification.

12.1.2 Computation Models

Among the most important aspects of domain definition is providing *models of computation*. The difficulty in heterogeneous specification is not understanding how values are calculated, but rather how values change over time. When interfacing an analog component with a digital component, difficulties arise when dealing with behaviors of interface items. Specifically, how does the analog view of an input or output map to the digital view it may be interfaced with? The difficulty is not understanding the values that items take, but rather understanding how those values change over time with respect to each other.

As an example, consider the definition of the time quantity in the state_based and continuous_time domains. In the continuous_time domain, the time type is most easily represented as a real number, and the current time as an element of that type:

```
state_type :: type is real;
s :: state_type;
```

In contrast, in the abstract state_based domain, the state type is left largely unspecified. We know that it is defined, but not what its characteristics are:

```
state_type :: type;
s :: state_type;
```

The distinction between state in continuous_time and state_based is only that state is constrained to be real. Thus, continuous_time can be defined by extending state_based to include this constraint.

Defining models of computation and communication in Rosetta is accomplished in domains by defining a point of reference and how that point of reference changes. We have seen how time types, values, and functions can be specified. The key is choosing types appropriately and defining operations over those values. The essence of model-of-computation definition is how and when items change value.

Rosetta provides a collection of built-in factions to assist developers in defining models of computation. The most important are the infix operator @ and the next function, present in all domains except static. Literally interpreted, the expression x@s means "the value of x with respect to s." Thus, @ is a mapping from an item name and a point of reference to a value, and serves as a dereferencing function. Defining the type of s says much about the model-of-computation. If s is real, then we are modeling a continuous system. In contrast, if s is natural, we are modeling a discrete system. More exotic points of references include surfaces and events. If s is a Cartesian coordinate, we are modeling values along a two-dimensional surface. Using this approach, Rosetta models for heat dissipation in semiconductors can be defined. If s is an event, we can model processes and traces in a manner similar to a constraint satisfaction problem (CSP) or the pi-calculus.

By sequencing states and examining their associated environments, the concept of execution or computation is defined. Rosetta provides the next function over the state type to define this sequencing: next maps a state to its next state and can be constrained by referencing any aspect of the facet state. Thus, if x@s defines the value of x in state s, then x@next(s) defines the value of x in the state (or states) following s. Rosetta defines a standard shorthand x' that is interpreted as x@next(s), or "x in the next state." Defining next appropriately supports defining various models of computation using the same common syntax. The catch is that one must understand the current domain to understand how state changes.

The properties of next, together with the state type, define how states are sequenced, providing a basis for a model of computation. For example, given an arbitrary state type, the constraint next(s)>s forces the next state to always be greater than the current state. This ensures that the state will always move forward and forces the state type to be infinite. In contrast, if the constraint is removed and the constraint #state in natural is added, the state space becomes finite and states can be revisited. From a quite simple starting point, sophisticated models of computation can be provided.

When a domain provides no new declarations, but adds constraints to units of semantics, it is informally referred to as a *model-of-computation* domain. Typically, model-of-computation domains are defined by extending a unit-of-semantics domain or another model-of-computation domain. Where unit-of-semantics domains provide a basic vocabulary, model-of-computation domains constrain that vocabulary to define specific mechanism for computing and sequencing values.

12.1.3 Engineering Domain Definitions

As any engineer knows, simply understanding mathematics is not sufficient for solving real-world engineering problems. Mathematics is a prerequisite, but the real work is done using domain-specific knowledge built from that mathematics. When using Kirchoff's Laws, an electrical engineer rarely thinks of the continuous time model used by those laws. The civil engineer does not think about the dynamics equations underlying bridge design. An engineer works with domain information that abstracts away the need to go all the way back to mathematics. Because these abstractions are based on sound mathematical principles, they are safely used in design.

Engineering design languages must provide this abstract, domain-specific layer or they are not useful. Languages such as VHDL and Verilog specifically address the needs of digital circuit designers. Matlab and Mathematica provide a basis for defining systems that includes collections of libraries for specific engineering domains such as control and signal processing. The UML software specification language provides a basic modeling paradigm with the ability to define profiles for specializing to various software modeling domains.

For example, in digital design the concept of a clock is fundamental to system design. The definitions of event, quiet, rising-edge, and falling-edge are common when observing a clock and should be defined in any domain used for digital system specification. Concrete functions are provided having the following form:

```
event(x::bit)::boolean where event(x') == x/=x'
quiet(x::bit)::boolean where quiet(x') == x=x'
rising(x::bit)::boolean is event(x) and x=1;
falling(x::bit)::boolean is event(x) and x=0;
```

These definitions differ from those provided by unit-of-semantics and model-of-computation domains. First, they reference units of semantics, but unlike model-of-computation domains they do not constrain them. The use of ticked names implicitly references state, but the reference is only to examine values in that state. Second, they declare, but unlike unit-of-semantics domains, they immediately define new functions and variables. Of course, these characteristics are heuristic — it is quite possible to define a new unit-of-semantics in an engineering domain. However, by avoiding this, specifications are easier to write and use.

When a domain adds to a model-of-computation domain new definitions that reference, but do not constrain, units of semantics, it is informally referred to as an *engineering* domain. Engineering domains are the workhorse of Rosetta specifications. Most specifications written by designers use an engineering domain to define a specific model for a system or component. Unlike unit-of-semantics and model-of-computation domains, engineering domains are easily written and are commonly added to the domain set. Any time Rosetta is used in a new engineering discipline, new component collections are required, or new abstractions are needed for operations on models, engineering domains are the best modeling approach.

12.2 The Standard Domains

Figure 12.1 shows the domain semi-lattice populated with a classical domain set. The null, static, and descendants of the state_based domain are the most heavily used domains and serve as examples in this section. The static and null domains are the only two domains defined as required elements in the base Rosetta system. Users may add new domains or completely replace the domain lattice if necessary for their specification activities.

12.2.1 The Null Domain

The null domain refers to the empty domain containing no computation model and no vocabulary. Using the null domain results in a system with only the base

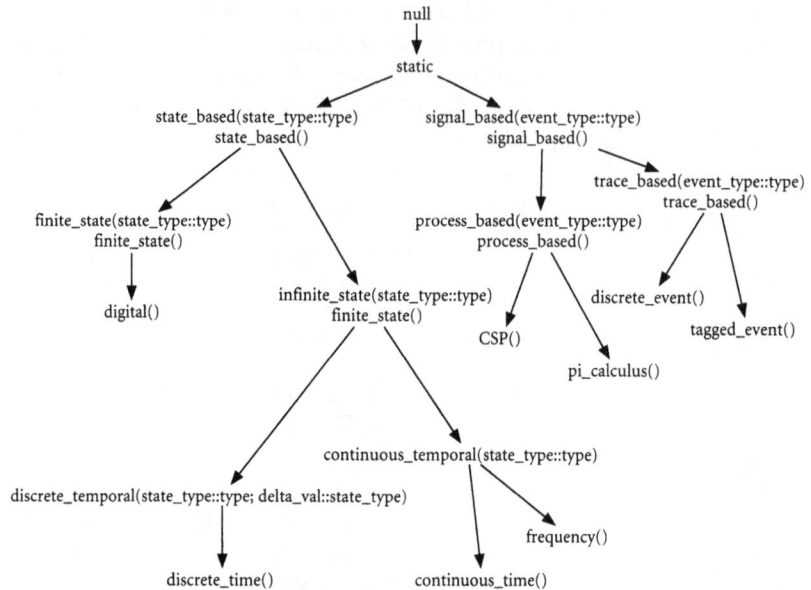

Figure 12.1   Base definition domains in the domain semi-lattice. Two domain signatures in the same node indicate different parameterizations of the same domain.

Rosetta semantics. It is included to provide a basis for defining domains that inherit nothing from other domains and should only be used by designers who want to start from scratch. The static domain that provides basic mathematics will use null as its type. Unless you intend to define a complete Rosetta domain semantics from scratch, there is little need to reference the null domain. There is no mechanism for constructing null from other domain definitions.

12.2.2 The Static Domain

The static domain provides a basic collection of types and functions in a computational model that has no built-in concept of state or time (Figure 12.2). The static domain is defined by using the prelude and unicode packages providing access to basic Rosetta definitions. The original names for the static domain

```
use rosetta.lang.prelude;
use rosetta.lang.unicode;
domain static()::null is
begin
end domain static;
```

Figure 12.2 Skeletal specification of the static domain.

were logic and math, reflecting the domain's intent of providing basic mathematical definitions. All predefined domains inherit from static and thus share the same mathematical basis. Symbols used to define computational models, @, next, and the state type are not visible in the static domain and cannot be used. Because static has no built-in state or time definitions, domains that inherit from it are free to define their own computational models.

The static domain is most frequently used when defining utility packages where model-of-computation is immaterial or harmful. Operations such as square root and function application, and data types such as integers and sequences, are the same regardless of computation model. The static domain provides common definitions used by all domains and thus all facets. The static domain is also used when defining structural definitions or performance requirements where models of computation are not used. The most important use of the static domain is as a basis for defining remaining domains. Because all domains share the static domain, all domains share the same mathematical structures and Rosetta support functions.

EXAMPLE 12.1

Power Constraint
Specifying a Static Power
Limit

The powerLimit facet specifies a limit on the value of an observed item, power. It simply states that the value of that parameter cannot exceed the statically defined value limit. Such a facet would be used as a constraint in a system where power is being consumed and should be limited. The static domain is used because the limit does not change.

```
facet powerLimit(power::real; limit::design real) :: static is
begin
  c: power =< limit;
end facet powerLimit;
```

When used in a system specification, the domain semi-lattice is typically used to transform the static powerLimit facet to a domain appropriate for composition, with a specification changing power values. The constant nature of the power limit is transformed into a constant limit using the target domain's computation model. By using the static domain, this specification can be moved virtually anywhere in the domain semi-lattice without sacrificing utility. However, the author of the power facet need not know anything about the target specification. This idea is fully explored in Chapter 15. ∎

EXAMPLE 12.2

Facet Containing Common
Function Definitions and
the Equivalent **package**
Specification

The addFns specification defines a pair of utility functions for defining addition operations in digital systems. The static domain is used because there is no concept of time when defining this type of pure mathematical function.

```
facet addFns() :: static is
  export all;
  bitSum(x,y,cin::bit) :: bit is x xor y xor cin;
  bitCarry(x,y,cin::bit) :: bit is
    (x and y) or (x and cin) or (y and cin);
begin
end facet digital;
```

This facet can be included in other facets and functions referenced using the dot notation. However, such utility specifications are so common in Rosetta that the **package** structure is included to define them. As a rule-of-thumb, whenever a facet is defined that includes declarations with no terms, a package should be used instead. Following is the equivalent definition using a package:

```
package addFns()::static is
  bitSum(x,y,cin::bit) :: bit is x xor y xor cin;
  bitCarry(x,y,cin::bit) :: bit is
    (x and y) or (x and cin) or (y and cin);
end package addFns;                                    ∎
```

12.2.3 The State-Based Domain

The state_based domain is a unit-of-semantics domain used to define systems that change state. The state_based domain extends the mathematical capabilities provided by the static domain to include the concept of state and change as traditionally defined. In the static domain, values associated with items cannot change because there is no concept of state. The state_based domain provides the most basic concept of change by introducing the next function and the state type described previously. Specifically, the state_based domain provides the basis for modeling the concepts of state and change by defining (i) an abstract state type, (ii) the current state, (iii) a next state function that derives the next state from the current state; (iv) and an event predicate that is true when an items Value changes from its Value in the previous state. Computations are defined by specifying preconditions on the current state and a relationship between the current and next states that defines computation (Figure 12.3).

```
domain state_based()::null is   domain state_based(state_type::type)::null is
  state :: type;                  state :: type is state_type;
  s::state;                       s::state;
  next(s::state)::state;          next(s::state)::state;
  event(X::top):: boolean;        event(X::top):: boolean;
begin                           begin
end domain state_based;         end domain state_based;
```

Figure 12.3 Partial specification of the parameterized and unparameterized state_based domains, showing the declaration of the state type, the current state, and the next state functions.

As a simple example, consider the following definition of a component that sorts its input:

```
facet sort(i::input sequence(integer);
           o::output sequence(integer))::state_based is
  sorted(i::sequence(integer))::boolean is
    forall(x::0,..((#i)-1) | i(x) =< i(x+1));
begin
  pre: true
  post: sorted(o') and ~o'=~i
end facet counter;
```

This facet defines requirements for a component that sorts its input without specifying a specific sorting algorithm. The state_based domain is used because we must define the output in the next state, but we do not know what the actual state is. Requirements defined by this component can easily be met by either a software or hardware component.

```
post: sorted(o') and ~o'=~i
```

The post term defines a post-condition for the component by defining properties on values in the next state. Specifically, o' refers to the value of o in the next state; post asserts that o' must be sorted and must be a permutation of the current input, i.

It is exceptionally important to recognize that the following term, similar to a C-like programming statement, is not correct:

```
next_s: o = sortFn(i);
```

Terms simply state things that are true. They are not executed and there is no notion of assignment. Although legal in C, where "=" is an assignment operator, in Rosetta this statement asserts that o=sortFn(i) is true. Looking at "=" as equality rather than assignment makes this statement inconsistent because we are asserting that the output is sorted in the same state in which we observe the input. This simply cannot hold. A good rule-of-thumb is to use the tick notation whenever the semantics of assignment is desired.

The key to using state_based domains is recognizing that an un-ticked label indicates the current state and that a ticked label indicates the next state. The state_based tick notation is defined based on the more fundamental state_based domain definitions of current state and the next state function. In reality, the notation x' is shorthand for the notation x@next(s), where (i) @ refers to the value of a label in a state, (ii) next defines the state following a given state, and (iii) s is the current state. Specifically:

$$x \stackrel{\text{def}}{=} x@s$$
$$x' \stackrel{\text{def}}{=} x@next(s)$$

The previously defined sort specification is equivalent to the following expansion:

```
facet sort(i::input sequence(integer);
            o::output sequence(integer))::state_based is
  sorted(i::sequence(integer))::boolean is
    forall(x::0,..((#i)-1) | i(x) =< i(x+1));
begin
  pre: true
  post: sorted(o@next(s)) and ~(o@next(s))=~(i@s)
end facet counter;
```

Here the tick notation is replaced by its definition using the next function, and undecorated variables are referenced in the current state. In the state_based domain, the dominant specification methodology is *axiomatic specification*. Thus, the primary use of @ is to refer to variables in the next state.

Statements of the form R(x,y@next(s)) or x∘y@next(s) are used to constrain the value of y in the next state, based on the value of x in the current state. This defines requirements for a computation without defining implementation method. When using the state_based domain, defining a relationship between the items in the current and next state should always be the specification goal. Revisiting the sort specification, the term labeled post defines the relationship between current and next state by defining a relationship between variables in those states. One of the advantages of the the state_based domain and Rosetta generally is that the next state need not be completely defined. If only partial requirements are available or desired, it is necessary only to specify what is known. At high levels of abstraction, such capabilities help the designer avoid over-specification during design exploration.

The state_based domain provides a special function, event, that compares the value of an item in the current and next states to determine if it has changed values. Specifically, event(clk) is true when clk changes values between the current and previous states. The event function supports definition of reactive systems that are triggered by changes in their inputs. A specific example of the event function in action is the definition of a rising clock edge:

```
rising(clk::bit)::boolean is event(clk) and clk=1;
```

In this definition, rising is true if event indicates a change in the value of clk and the resulting of the change is a value of 1.

What is happening in the state_based domain is that the model-of-computation is being exposed in the specification. Languages such as C, VHDL, and Lisp have a built in computational model that pervades all definitions. In Rosetta, the model is exposed, allowing integration of specifications in different domains. The price paid for this freedom is the need to think about the model-of-computation in the specification process. Rosetta tries to abstract away as much detail as possible, but some will always remain.

To summarize, the state_based domain should be used whenever a system-level description of component or system is needed. At early design stages, when working at high levels of abstraction, the state_based domain provides a mechanism for describing state transformations without unnecessary details. The state-based domain should not be used when details such as timing are involved in the specification. Furthermore, the state_based domain provides no automatic mechanism for composing component states when developing structural models.

EXAMPLE 12.3

Register Specification Using the state_based Domain

In the Introduction we explored the definition of a register using the discrete_time domain, examining specification composition and the use of heterogeneous domains. Here we examine a more robust specification in the state_based domain that uses a universally quantified parameter to allow arbitrary register word width.

```
facet reg
   [size::natural](x::input word(size); z::output word(size);
    rst,le,clk::input bit)::state_based is
 v :: word(size);
  init(s::natural)::bitvector is
    if s=0 then [] else cons(0,init(s-1)) end if;
begin
  sup: v' = if reset=1 then init(size) else
              if clk=1 and event(clk) and le=1
                then x
                else v
              end if;
            end if;
  zup: z = v;
end facet reg;
```

Specification itself defines a state update term (sup) and an output update term (zup). The state update term references the facet's state variable, v, in the next state using the tick notation, v'. If the clock is rising and load enable is set, then the register loads the input value, x. If any of these conditions is false, the register's value remains unchanged. An outer condition checks for a reset value. By using v' to reference the next state value, the state_based specification vocabulary is used.

The output update term asserts that the component output, z, is equal to v in all states. Thus, the output is the value of the stored value, v. There is no condition on this equivalence, so the condition must always be true.

The universally quantified parameter, size, represents the length of the word being stored. It is specified as a universally quantified parameter to allow its value to be inferred at elaboration time. Like VHDL, Rosetta can determine the width of word type variables by propagating widths through component interconnections. Rosetta differs because the universally quantified parameters need not be specified and can be specified explicitly. If a universally quantified parameter's value cannot be determined, it may simply indicate early design stages, not an error. If the user does not want tools to try and infer values, universally quantified parameters can be explicitly instantiated in the same manner as traditional facet parameters. ∎

EXAMPLE 12.4

Modeling Activity-based Power Consumption Using the state_based Domain in a Facet and in a New Engineering Domain

The powerConsumption facet is a state-based specification describing power consumption. The technique used in this model is a trivial form of *activity-based power modeling*. When signals change state, the system consumes power. In this model, leakage and switch are design parameters representing the nominal power consumption and power consumed by a state change, respectively. Total power consumed is measured by summing power consumed over sequences of state changes.

```
facet powerConsumption(o::output top;
        leakage,switch::design real)::state_based is
  export power;
  power::real;
begin
  power' = power + leakage + if event(o) then switch else 0 end if;
end facet powerConsumption;
```

The model for this specification uses the state_based domain because it is the most abstract domain where state change occurs. The specifics of state change are not required by the activity-based model, thus the state_based domain is an appropriate choice. The power value after each state change is the power value before the state change plus leakage and switch if the value observed changes.

This model is naive, but can easily be expanded to cover larger systems with additional inputs and outputs. Its abstract nature allows it to be specialized for domains that extend state_based. However, knowing more about the state change allows specifying more power consumption details. Defining a power domain based on state_based would allow inheriting these basic power consumption concepts. Such a domain would have the following form:

```
domain power_activity(leakage,switch::design real)::state_based is
  powerUpdate(o::top)::real is
    leakage + if event(o) then switch else 0 end if;
begin
end domain powerConsumption;
```

The following specification redefines the original power consumption specification using this powerConsumption domain. Again, the model is simple but demonstrates the definition of one domain inheriting from another. Specifically, the power_activity domain is an engineering domain that defines vocabulary for activity-based power consumption modeling.

```
facet powerConsumption(o::output) :: power_activity(leak,switch) is
  export power;
  power :: real;
begin
  power' = power + powerUpdate(o);
end facet powerConsumption;
```

12.2.4 The Finite State Domain

The finite_state domain is a model-of-computation domain that provides a mechanism for defining systems whose state space is known to be finite and can be listed or represented using a type. Systems such as RTL components, controllers, and protocols are defined using the finite_state as a basis domain. The finite_state domain extends the state_based domain, thus all definitions from state_based remain valid in the new definition. Specifically, next, @, and tick retain their original definitions. The only addition is the constraint that state must be a finite type (Figure 12.4).

```
domain finite_state(state_type :: type)::state_based is
  state :: type is state_type;
begin
  #state_type in natural;
end domain finite_state;
```

Figure 12.4 Partial specification of the finite_state domain. Definition of the state type and finite constraint.

The finite_state domain extends the state_based domain by adding a constraint on the size of the state_type value. The state_type parameter specifies the collection of items used to represent the state and must be a type. The declaration of state gives it the value specified by state_type.

Unlike the state_based domain, the finite_state domain has constraints on states. The new term defines the size of state_type to be a natural. Because all individual naturals are finite, this assures that the state set must also be finite. This constraint must be satisfied by any value specified for state_type. Care must be taken to assure that the size of state_type can be calculated or, minimally, is not known to be infinite.

Consider an example counter:

```
facet counter(clk::in bit; c::output natural)::finite_state(0..,3) is
  rising(x::bit)::boolean is event(x) and x=1;
  next(s::state)::state is
    if rising(clk) then
                    if s =< 2 then s+1 else 0 end if
                else s end if;
  begin
    next_s: s' = next(s);
    out: c = s;
  end facet counter;
```

In this counter, the state space is explicitly defined as the set containing 0 through 3 by using the state_based domain's parameter. The new definition of the state type is the set {0..,3}. Because #state=4, the state space is finite, causing no inconsistency with the requirement that the finite_state domain have a finite state type. Instead of defining the next function in terms of properties of the next state, it is explicitly defined as modulo 4 addition on the current state when the clock is rising. This is quite different than previous state_based specifications where the actual value of the next state was not defined.

The terms assert that the next state is calculated by next and the next output is the next state. If the clock does not rise, the current state is maintained and thus the current output. The term:

```
next_s: s' = next(s);
```

performs the state update, while the term:

```
out: c = s;
```

performs output. If the state does not change, then neither will the output.

Like state_based specification, finite_state specification does not require the next function to be completely defined, although it is more typical for this to occur in the finite_state domain, as it is less abstract than the state_based domain. It should be noted that the state type and the next state function are defined in the finite_state domain definition and need not be redeclared.

The finite_state domain is only useful when defining systems known to have finite state types. For example, whenever a sequential machine is desired, the finite_state domain is the appropriate specification model. Typically, the elements of the state type are specified by extension or comprehension over another bunch to assure that the state type is finite. Most RTL style specifications can be expressed using the finite_state domain, if desired.

The finite_state domain should not be used when the state type is not known to be finite or is not known at all. Of particular note is that finite_state specifications should not be used when timing information is specified as a part of function. In such circumstances, the set of possible states is almost always infinite, because time has no upper bound. An interesting challenge in heterogeneous specification is understanding when an infinite system can be represented using a finite number of states by abstracting away the specifics of time.

EXAMPLE 12.5

Alternative Definition for the counter Specification Using a Term to Define the next Function

The following counter facet definition is an alternative for the earlier definition. Rather than using the constant function syntax, a term defines the next function's value:

```
facet counter(clk::in bit; c::output real)::finite_state(top) is
    rising(x::bit)::boolean is event(x) and x=1;
begin
    next_state: forall(s::state | next(s)=if rising(clk)
                                          then if s =< 2 then
                                                    s+1 else 0
                                               end if
                                          else s end if;
    next_s: s' = next(s);
    out: c = s;
end facet counter;
```

Semantically, this definition is the same as the original definition. The special syntax used for function definition in the original discussion is easier for tools to recognize, thus the earlier syntax is preferred. ∎

EXAMPLE 12.6

Synchronous, Finite Buffer That Contains Data of Some Arbitrary Type Message

The buffer specification represents a finite buffer that is updated when a new input arrives, and that outputs when it is not blocked and the clock rises. A package is used to define a type representing the contents of the buffer as a finite sequence of some message. The finiteSequence function returns a type consisting of all message sequences of length less than or equal to n. This is identical to the word specification except that all elements of the word type have exactly the specified length. This is not appropriate here because the buffer changes size when messages arrive and are processed. The heart of the specification is the n1 term that constructs the next state.

Given some state s, the next state is constructed by knowing the status of the in, block signal, and clk. If a new message is to be output, then the message is removed from the front of the list. At the same time, a message could be arriving. If so, the new message is added to the end of the buffer.

```
package bufferType()::static is
  message :: type;
  finiteSequence(n::natural)::subtype(sequence(message)) is
      s::sequence(message) | #s =< n ;
end package bufferType;

use bufferType();
facet buffer(in::input message;
            out::output message;
            block,clk::input bit)::finite_state(finiteSequence(8))) is
begin
  n1: next(s) = if not(block) and event(clk) then cdr(s) else s end if
               \& if event(in) and #s < n then [in] else [] end if;
   o1: out' = if not(block) and event(clk) then car(s) else out end if;
end facet buffer;
```

12.2.5 The Infinite State Domain

Like the `finite_state` domain, `infinite_state` is a model-of-computation domain that extends the `state_based` domain by restricting the state definition. Instead of requiring that the state type be finite, the `infinite_state` domain implicitly makes the state type infinite by enforcing a total ordering on states (Figure 12.5). The term `next(s)>s` states that the next state is always greater than the current state, effectively eliminating loops in the state transition function. Mathematically, the definition of `next` makes the state type a chain. If a state were to appear twice it would violate the ordering property. The `infinite_state` domain is useful when defining systems where states are ordered and potentially infinite numbers of states exist. Representing a discrete event system is an appropriate application of the `infinite_state` domain.

The `infinite_state` domain exists principally to serve as a common domain for defining systems with time. It should not be used when the specifics of timing and time measurement are known. The `discrete_time` or `continuous_time` domains are chosen over the `infinite_state` domain in most modeling situations. Both domains inherit properties from the `infinite_state` domain but provide additional details about the time type.

```
domain infinite_state(state_type :: type)::state_based is
  state :: type is state_type;
begin
  next(s)>s;
end domain infinite_state;
```

Figure 12.5 Partial definition of the `infinite_state` domain, showing definition of the state type and the infinite state constraint.

The dataLogger facet is an example of using an infinite state specification to model potentially infinite input and output data sequences. The state of this specification is a pair of real values, thus a more concrete temporal domain is not appropriate. Each time the clock rises, dataLogger samples its inputs and adds these values to the existing x and y values maintained by the state. A reset signal may be used to reset the state.

```
package pairData()::static is
  pair :: type is data
    pair(x::real,y::real)::ispair;
  end data;
end package pairData;

use pairData;
facet dataLogger(dx,dy::input real;
                 clk,rst::input bit;
                 o::output pair)::infinite_state(pair) is
begin
  xupd: x(s') = if rst = 1
                  then 0.0
                  else x + if rising(clk) then dx else 0.0 end if
                end if;
  yupd: y(s') = if rst = 1
                  then 0.0
                  else x + if rising(clk) then dy else 0.0 end if
                end if;
  out: o=s;
end facet dataLogger;
```

This specification is also interesting because the update of the state value is not directly defined. Specifically, there is no point where s' is equated with any new value. Instead, the x and y values for the next state are constrained. We are defining the values of observations in the next state rather than the next state directly. This is an exceptionally common specification technique that is particularly useful when the implementation of state is not completely known or the characteristics of a state cannot be fully specified. ■

12.2.6 Discrete Time Domain

The discrete_temporal and discrete_time domains are model-of-computation domains that define models where time is observed in discrete time intervals. Both domains inherit definitions from the infinite_state domain, including the total order on the state type. The discrete_temporal domain is parameterized over the time type and time increment, defining only discrete time restrictions. The discrete_time domain specifies the time type as natural and the time increment as 1.

The discrete_temporal domain adds restrictions to the infinite_state domain to ensure that the state type is discrete and time values increase uniformly from one state to the next. The discrete_temporal domain is parameterized over both the time type and increment. The next state function is defined

```
domain discrete_temporal(state_type :: type;
                         delta_val :: state_type)::infinite_state is
  state :: type is state_type;
  delta :: state is delta_val;
  next(t::state)::state is t+delta;
begin
  #state_type = #natural;
end domain discrete_temporal;

domain discrete_time()::infinite_state is discrete_temporal(natural,1);
```

Figure 12.6 Partial specification of the discrete_temporal and discrete_time domains.

as next(t)=t+delta and, following from previous domain definitions, x@t is the value of x at time t and x' is equivalent to x@next(t) (Figure 12.6).

Discreteness of state_type is enforced by requiring its cardinality to be the same as natural. The only restrictions on delta are made indirectly by the definition of next. The next function defines the next state as the current state plus delta. The infinite_state restriction on next(t) ensures that delta cannot be zero and next must increase the state value.

The discrete_time domain is a specific instantiation of the discrete_temporal domain. It specifies the time type as natural, the delta value as 1. Specifically, in the discrete_time domain, time is a natural number denoted by t and discrete time increment is 1. This domain provides a convenient standard definition for discrete time Rosetta specifications.

Specifications are written in the discrete time domain in the same fashion as the infinite and finite state domains. The additional semantic information is the association of each state with a specific time value. Thus, the term:

```
t1: x' = f(x)
```

constrains the value of x at time t+delta to be the value of f(x) in the current state. This specification style is common and reflects the general syntax and semantics of a VHDL signal assignment. Specifically, if delta were specified in femptoseconds, the discrete time model would be an excellent low-level model of VHDL signal assignment. As such, the discrete time domain is frequently used when looking at systems where periodic clocks are appropriate. The notation:

```
x@(t+(n*delta)) = f(x)
```

provides a mechanism for looking several discrete time units into the future. Such mechanisms are useful when defining delays in digital circuits.

The discrete_time domain should be used when modeling requires a timing model that utilizes fixed, discrete time units. This domain provides modeling capabilities that map nicely onto digital logic and communications systems, where timing issues are well understood. The discrete_time domain should not be used when no fixed timing constraints are known. In such situations, the

EXAMPLE **12.8**

Digital Systems
Specifications

`infinite_state` or `state_based` domains may be more appropriate and will help avoid over-specification.

An interesting application of discrete time modeling is the development of digital system specifications. Traditionally, digital designs specify a clock and then monitor that clock for events (rising and falling edges). The `discrete_time` domain can be used to model such systems without complicating the definition with clock management. For example, a VHDL process that is rising-edge triggered might specify a wait condition as:

```
wait on clk;
```

followed by a check to see if the clock is high. Rosetta abstracts this detail out in the discrete time domain by associating a clock period with the discrete time interval. Specifically, by specifying system values with respect to state, values are implicitly specified with respect to clock cycles. A specification of a simple register has the following form:

```
facet register(x::in bitvector; z::output bitvector;
               e::in bit; r::in bit)::discrete_time is
  s::bitvector;
  choose(b0,b1::bitvector; e,r::bit)::bitvector is
     if r=1 then 0
        else if e=1 then b1 else b0 end if
     end if;
begin
  t1: s'=choose(s,x,e,r);
  t2: z'=choose(z,s,e,r);
end facet register;
```

This specification defines a cycle-based model of the register. The function `choose` is defined to select a value based on enable and reset inputs. This function is not necessary, but simplifies the specification by abstracting out the selection properties of the enable and reset signals. The terms specify the next value of the stored value and the output value. Because the specification assumes that all changes occur on the initiation of a new clock cycle, the clock can be left out of the specification.

It is important to note that the reset signal for this register is synchronous. Because changes that do not occur with respect to a clock signal cannot be modeled, including an asynchronous clock is not possible. Although the abstract specification is more useful for many purposes, the trade-off is information content.   ∎

12.2.7 The Continuous Time Domain

The `continuous_time` domain is a model-of-computation domain that provides a mechanism for defining temporal specifications using a real-valued representation of time. Unlike discrete time specifications, continuous time specifications allow reference to any specific time value by treating time as a `real` value.

The continuous_time domain provides a state type representing time that is uncountably infinite (Figure 12.7). This differs from the discrete_time domain, where time values are restricted to a countably infinite set. In virtually all specifications, this time type will simply be real.

```
domain continuous_temporal(state_type :: type)::infinite_state is
  state :: type is state_type;
  delta :: state;
  next(s::state)::state is lim(s+delta,delta,0);
  tderiv(x::label) = deriv(x,s);
  tantideriv(x::label) = antideriv(x,s);
  tinteg(x::label,u::real,l::real,c::real) = integ(x,s,u,l,c);
begin
end domain continuous_temporal;

domain continuous_time::infinite_state is continuous_temporal(real);
```

Figure 12.7 Partial specification of the continuous time domain.

The next function is more difficult, as it refers to the instantaneous next state. Specifically, what does x@next(t) or x' mean when time is continuous? In discrete time, we defined next(t) = t+delta, where delta is a discrete time increment. In continuous_time, we want delta to approach zero. Specifically:

next(t::state) = $\lim_{t\to 0}$ t+delta

defines the next state to be the instantaneous next state. Using the Rosetta built-in limit function, the formal definition becomes:

next(t::state) = lim(t,delta)

Thus, x@next(t) and x' define the the value of x in the next instantaneous state. Using this definition of next, we can define transfer functions and provide time derivative, antiderivative, and integral functions for general use.

The time derivative, or rate of instantaneous change associated with an item x, is defined as x'deriv(t) or by viewing x as a function of time. An nth order time derivative can be referenced by recursive application of deriv. The second derivative is defined as x'deriv(t)'deriv(t), the third derivative as x'deriv(t)'deriv(t)'deriv(t), and so forth.

Remember that in any state-based specification system, an item x is defined with respect to some observation reference. Specifically, x expands to x@t, where "@" is making an observation with respect to time. We can express this using a more traditional functional notation as x(t) and treat the variable x as a traditional function over time. With this, we can define a traditional time derivative as:

$$\frac{dx}{dt} = \frac{d}{dt}x(t)$$

Using Rosetta syntax and the derivative function this becomes:

```
x'dot = x'deriv(t)
```

The time derivative of x is simply the derivative of x with respect to t. The dot ticked expression is provided as a shorthand, making time implicit in the derivative.

The indefinite integral with respect to s is defined similarly as x'tantideriv and behaves similarly when x is an item in the continuous_time domain. Note that the antiderivative with respect to time assumes an integration constant of zero. Making the integration constant different is a simple matter of adding or subtracting a real value from the indefinite integral.

The definite integral with respect to time is provided as x'tinteg(1,u) and is defined in the canonical fashion as:

$$\text{x'tinteg(1,u)} \stackrel{\text{def}}{=} \text{x'tantideriv(u) - x'tantideriv(1)}$$

The continuous_time domain should be used when dealing with analog electronic systems or any other system referencing continuous time. Such systems include mechanical, optical, biological, and chemical systems in which discrete time has no real meaning; continuous_time should not be used when the time reference is known to be discrete. Although such systems can be modeled in the continuous_time domain, the various discrete time domains capture the discrete nature of time. Using the continuous_time domain, this must be inferred during verification.

EXAMPLE 12.9

Two Facets Alternative models of an Analog-to-Digital Converter in the continuous_time Domain

The following specification for a simple analog-to-digital converter seems to be quite correct. The terms specify cases for the two possible current output values. The first case checks if the clock is rising and the output is high, while the second checks if the output is low. If either case is met then the output is driven appropriately, based on the input. If neither case is met, the **else** clauses for both terms assert the output should remain the same.

Unfortunately, if the condition for one term is met, the condition for the other cannot be met. This is due to the type associated with the output value. In this case, one term may constrain the output to be high while the other is low. This is an excellent example of an inconsistent specification. Cases like this, where the same item can be constrained inconsistently, must be avoided and are difficult to automatically detect.

```
facet a2d(i::input real; o::output bit; clk::input bit)::continuous_time is
begin
  outh: if rising(clk) and o=1
          then o'= if i<0.3 then 0 else 1 end if
          else o'=o
        end if;
  outl: if rising(clk) and o=0
          then o'= if i>0.7 then 1 else 0 end if
          else o'=o
        end if;
end facet a2d;
```

The next specification avoids the problem in the first specification by using a single **if** expression that ensures only one assertion is made about the output. Here, if the clock is rising, the input and current output are checked to determine the next output value. The **if** expression guarantees mutual exclusivity of assertions on the output, avoiding the inconsistency in the previous specification.

```
facet a2d(i::input real; o::output bit; clk::input bit)::continuous_time is
begin
  out:  o'=if rising(clk)
            then if o=1 and i<0.3 then 0
                 elseif o=0 and i>0.7 then 1
                 else o end if;
            else o
          end if;
end facet a2d;
```

The continuous_time domain is used for this specification because its input is an analog signal. Its output is binary valued, but it is continuous as well. It is the responsibility of the component using the a2d output to lift it into the digital domain using translator functions.   ∎

EXAMPLE 12.10

Passive Lowpass, Highpass, and Bandpass Filters Defined Using the continuous_time Domain

The lowpass and highpass facets are time domain definitions of simple RC filter circuits. Parameterized over r and c, they represent the structure of classical, first-order passive filters. In both cases the next output is determined by the current input voltage, r and c. Recall that in the continuous_time domain, the next output is the instantaneous next output. Thus, we are defining the transfer functions for these simple circuits.

```
facet lowpass(in::input real; out::output real;
              r,c::design real)::continuous_time is
begin
  filter: out' = in*(1/sqrt(1+(r*c*in'deriv(t))^2))
end facet lowpass;

facet highpass(in::input real; out::output real;
               r,c::design real)::continuous_time is
begin
  filter: out' = in*(r*c*in'deriv(t)/sqrt(1+(r*c*in'deriv(t))^2)
end facet highpass;
```

Virtually any time domain system can be written in the same manner and composed to define more complex devices. For example, the following structural code uses lowpass and highpass filters to define a bandpass filter:

```
facet bandpass(in::input real; out::output real;
               rlow,clow,rhigh,chigh::design real)::continuous_time is
  x::real;
begin
  lpf: lowpass(in,x,rlow,clow);
  hpf: highpass(x,out,rhigh,chigh);
end facet bandpass;
```

By placing the lowpass and highpass filters in sequence, a passive RC bandpass filter is defined.   ∎

12.2.8 The Frequency Domain

Another continuous domain used pervasively in systems design is the frequency domain. We define the `frequency` domain using the `continuous_temporal` as a kind of template. We define the `frequency` domain as:

```
frequency :: infinite_state is continuous_temporal(real);
```

One should notice quickly that the `frequency` domain and the `continuous_time` domain are identical, with the exception that the continuous reference value is viewed as frequency rather than as time. This is to be expected, as all operations performed over time domain objects can also be performed in the frequency domain. They frequently have completely different interpretations, but their basic definitions are the same.

Another aspect worth mentioning is the existence of a functor between the frequency and continuous time domains. It is possible to move specifications between the time and frequency domains in mathematics using the Fourier transform. Similarly, the Fourier transform and its inverse define functors between the frequency and time domains. Functors and their uses are defined fully in Chapter 15, however, the utility of the Fourier transfer illustrates the value of having functors and interactions.

EXAMPLE 12.11

Passive RC Filter
Specifications Defined in
the frequency Domain

The `lowpass` and `highpass` facets defined below are frequency domain definitions for the RC filter circuits from the previous example.

```
facet lowpass(in::input real; out::output real;
        r,c::design real)::frequency is
begin
  filter: out' = in/(1+r*c*w)
end facet lowpass;

facet highpass(in::input real; out::output real;
        r,c::design real)::frequency is
begin
  filter: out' = in*r*c*w/(1+r*c*w)
end facet highpass;
```

The domain for these facets is frequency, not an abstraction of time. Variables are referenced to frequency, referred to as w. The result is that **in** and out are functions of w. The same is true for r and c, but they are of kind **design** requiring that they do not change with respect to frequency.

```
facet bandpass(in::input real; out::output real;
                rlow,clow,rhigh,chigh::design real)::frequency is
  x::real;
begin
  lpf: lowpass(in,x,rlow,clow);
  hpf: highpass(x,out,rhigh,chigh);
end facet bandpass;
```

By placing the lowpass and highpass filters in sequence, a passive RC bandpass filter is defined in the same manner as before. ∎

12.3 Domains and Facet Types

It is no coincidence that when a new facet is defined, its domain is identified using the Rosetta type indicator, "::". Every facet is considered to be an element of a type identified by its domain and a subtype of every domain its domain inherits from. Using this concept, we define how a facet type is identified by a domain and how domains form a lattice of specifications.

12.3.1 Domains as Types

Each time a facet is defined, it must be associated with a modeling domain. The facet identifies its domain by including it immediately following its parameter list, in a manner similar to defining a function's range:

```
facet interface adder(x,y,ci :: input bit; z :: output bit) :: digital is
end facet interface adder;
```

As we have seen, the specified domain provides a basis for building the facet and the type the facet is associated with. Each facet definition extends the domain it includes by adding declarations and terms to the domain definition. Thus, the domain provides a base for facet specification while additional definitions provided by the facet extend the domain to include system-specific properties. The term extension is used because the properties expressed in the basic domain definition cannot be eliminated, only enhanced. For example, if working in the discrete_time domain, one cannot change the properties of the next command other than by adding new properties. The basic definition of next must be maintained.

In addition to providing a modeling vocabulary, the domain associated with a facet defines the facet's type. A domain's associated type is defined as every consistent extension of that domain. This specifically includes all facets and domains created by consistently extending the domain. In this way, the domain serves as a specification of all elements of its associated type. When we say:

```
facet adder(x,y,ci :: input bit; z,co :: output bit)::digital is ...
```

the "::" operator plays the same role it always has by identifying the facet type. adder is of type digital and can be treated like a value in the same way as numbers, strings, sequences, or any other Rosetta value. This allows definitions of functions over facets that enable defining architectures and facet composition among other useful constructions.

We can also say:

```
halfadder(x,y :: input bit; z,co :: output bit)::digital is adder (x,y,0,z,co);
```

in a manner quite similar to function definition. Here we are saying that a halfadder is of type digital and is an adder with the carry-in value set to 0. The type is used to assure that the newly created facet value associated with halfadder is the same type as the type of halfadder. The key is that facet types are no different than other types. They play the same role when defining new variables and values associated with a given domain.

As an extreme example, one can define facets or functions using facet type parameters. An excellent example is the Fourier transform functor that is simply a function with a domain of continuous_time facets and range of frequency facets:

```
fourier(f::continuous_time) :: frequency is
  make_anonymous_facet(
    make_facet_signature(parameters(facet_signature(f)),'frequency'),
    exports(f),
    declarations(f),
    << infix_binary_operation
      z =
      'right_argument((terms(f)(0)))' *
      exp(-j*w*t)'tinteg(-infinity,infinity)/sqrt(2*pi)'
      >>);
```

where f is a facet defining a time domain transfer function that defines an equivalence between a dependent time domain parameter z representing the facet output and an independent time domain variable x.

Facets can be similarly parameterized, providing a way for writing general-purpose architectures. The signature for an architecture that connects digital components in sequence would be:

```
facet digital_sequential(f1,f2 :: static digital;
                         i::input bit; o::output bit) :: digital
  x::bit;
begin
  f1(i,x);
  f2(x,o);
end facet digital_sequential;
```

The sequential facet accepts two digital components and passes data through f1 and on to f2. This facet can be used to instantiate components in an architecture:

```
neg_neg(i::input bit,o::output bit) :: digital is
  digital_sequential(inverter,inverter,i,o);
```

The architecture facet can utilize universally quantified parameters to provide an even more general architecture. Specifically, the types of the inner and outer components need not be specified directly, but could be inferred from the component's usage:

```
facet sequential
   [facet_T :: subtype(static);
    T :: type]
   (f1,f2 :: static facet_T;
    i::input T; o::output T) :: facet_T is
 x::T;
begin
 f1(i,x);
 f2(x,o);
end facet sequential;
```

The instantiation with inverter components is identical:

```
neg_neg(i::input bit,o::ouput bit) :: digital is
   sequential(inverter,inverter,i,o);
```

Type inference is used to determine facet_T and T. In this example, the inference is quite simple. The type of neg_neg is digital, thus facet_T must also be digital. This value satisfies the subtype constraint from declaration of facet_T and is also the type of inverter. Inferring T is equally simple. The parameters of neg_neg, i and o, are both of type bit, implying that T is bit and thus x is of type bit.

We will see further examples of such advanced abstraction techniques as our specifications become more general and reusable. For now, it is enough to understand that facets are Rosetta values and that, like any other value, they have associated types. We can define facet variables, parameterize functions over facets, and parameterize facets over other facets. Anything that can be done with any other item can also be done with facets and facet types.

12.3.2 **The Domain Semi-Lattice**

Domains are themselves facet structures and, like other facets, must identify the domain they extend. The only exception to this rule is the null domain that by definition is the initial domain containing nothing. Thus, domains themselves are elements of types and identify a type in their definition. For example, the finite_state domain is defined from the state_based domain as follows:

```
domain finite_state(state_type :: type)::state_based(state_type) is
begin
  #state_type in posint;
end domain finite_state;
```

The finite_state domain, is simply a state_based domain, where the cardinality of the state_type is a positive integer. As all positive integers are finite and non-zero, the size of state_type must be finite and non-zero.

It may seem odd that types themselves seem to have types, but it is quite common in object-oriented languages, where a class extends another class. Both can be used as types for new objects and one clearly extends the other. It also follows from the definition of a domain type. A domain's type is all possible consistent extensions of that domain. Any domain extending another domain satisfies this condition.

Using this extension or inheritance relationship, Rosetta defines the *domain semi-lattice* as a lattice structure of domain definitions. The structure is a semi-lattice, where the homomorphism defined by domain extension is the partial order. Specifically:

$$G \Rightarrow F \stackrel{\text{def}}{=} F \sqsubseteq G$$

If F :: G then $G \Rightarrow F$ and $F \sqsubseteq G$. Using this definition, the domain semi-lattice is a join semi-lattice whose top is null. If the inconsistent facet is introduced as the bottom value, the the semi-lattice becomes a full lattice.

Although the mathematics of the semi-lattice is important for Rosetta's semantics, the critical issue for specification is the relationship between domains. From this relationship, abstraction and concretization functions are defined that allow moving specification information up and down the semi-lattice. The existence of a Galois Connection in association with the semi-lattice assures the soundness of the transformations. When new domains are added to the semi-lattice, abstraction and concretization are defined, allowing facets written in the new domain to interact with all existing facets. As we shall see, this makes Rosetta's domain system exceptionally powerful for performing system-level design abstractions and transformations.

Reflection

Reflection, the ability to treat language constructs as data, is implemented in Rosetta using *abstract syntax structures*, *template expressions*, and *interpretation functions*. Abstract syntax structures, or simply AST structures, are Rosetta data types that internally represent the abstract syntax of Rosetta expressions. These structures are defined as Rosetta data types and can be manipulated like any other data type. Specifically, constructors, recognizers, and observers are defined for each AST structure to support creating and observing Rosetta specifications. AST structures are in every wadomainy Rosetta data structures and are defined in the standard `rosetta.lang.reflect` library.

Template expressions define syntactic structures that are parsed into their equivalent Rosetta AST structure during elaboration. Template expressions include concrete syntax that is parsed, as well as escaped functions that are called to directly generate AST structures. Although template expressions are not a semantic necessity, they simplify creating new AST structures by providing a representation syntax that resembles the AST construct being created.

Interpretation functions perform type inference, evaluation, and denotation of AST structures. The `value` function evaluates an AST structure in an environment attempting to reduce it to its simplest form. The `typeof` function infers the type of an AST structure given its declaration context. The function `phi` maps an AST structure to its semantic value. Finally, the `denotes` function denotes the value of an AST construct in Rosetta's co-algebraic semantics.

13.1 Template Expressions and AST Structures

Using AST structures, Rosetta specifications can refer to and create new specification elements. The basic syntax for Rosetta template expressions is:

> << *ast arg* >>

where *ast* identifies the type of AST structure created by the template and *arg* is the actual template. Such expressions create abstract syntax elements by identifying the type of an AST element and a template that is parsed into a structure of that type. The first element of a template expression, *ast*, identifies the data type for the resulting AST structure. For example, `expression` is the type representing a general Rosetta expression, while `ticked_operation` is the type representing a ticked expression. Thus, a template of the form:

```
<< expression e >>
```

will be parsed into an instance of the `expression` AST representing *e*. If *e* cannot be parsed into an expression, then an error results.

The second element of the template, *arg*, is the syntactic element to be parsed. For example:

```
<< expression x + 2 >>
```

creates the abstract syntax for an `expression` abstract syntax element representing x+2. The Rosetta elaboration system converts the template into a series of calls to abstract syntax data constructors. When evaluated, these constructors form the actual AST structure. For <<expression x+2>>, the resulting constructors have the following form:

```
make_infix_binary_operation (
   rosetta.lang.reflect.make_label ( "x" ), plus_token, 2 )
```

The constructor `make_label` creates a label named x, and `plus_token` is the parse token associated with the + operation. Finally, the constant 2 is the second order of the binary expression. The function `make_infix_binary_operation` creates the abstract syntax element from the individual abstract syntax elements.

Interpretation of template expressions can be escaped by enclosing syntactic elements in back quotes. Specifically, `'signature(n)'` indicates that the function instance `signature(n)` should be included without expansion. For example:

```
<< function_type <*'signature(n)'*> >>
```

expands to:

```
make_function_type ( ( signature(n) ) :: function_signature )
```

The function `signature(n)` is not evaluated until after the elaboration occurs, allowing n to be instantiated with an evaluation-time item. The label n may refer to a function declaration AST or other construct from which a function signature can be formed. When `make_function_type` is evaluated, the signature of n becomes the signature of the function type. One use for this function is extracting the function type from a constant function declaration.

EXAMPLE 13.1

Combining a Facet
Interface AST and Facet
Body AST to Create a
Complete Facet
Declaration

The following example uses Rosetta reflective AST functions to create a complete facet declaration from an interface and body. The function first checks to assure that the interface and body have the same names. If they do, the `make_complete_facet_declaration` constructor is used to assemble complete declaration. All elements of the declaration come from either the interface or body, except declarations. Both interface and body can have declarations, thus they are appended and used as the declarations value.

```
combine_elements(fi::facet_interface_declaration;
        fb::facet_body_declaration)::complete_facet_declaration is
    if facet_label(fi) = facet_label(fb) then
      make_complete_facet_declaration(
        facet_label(fi),
        facet_signature(fi),
        exports(fi),
        declarations(fi)&declarations(fb),
        terms(fb))
      else bottom
    end if;
```

A template expression can be used to accomplish the same declaration. However, because there is no real concrete syntax involved in new declaration, there is little benefit from doing so. To illustrate this option, the same function using template expressions has the form:

```
combine_elements(fi::facet_interface_declaration;
        fb::facet_body_declaration)::complete_facet_declaration is
    if facet_label(fi) = facet_label(fb) then
      << complete_facet_declaration
        facet 'facet_label(fi)' 'facet_signature(fi)' is
          'exports(fi)'
          'declarations(fi)&declarations(fb)'
        begin
          'terms(fb)'
        end facet 'facet_label(fi)' >>
      else bottom
    end if;
```

∎

13.2 Interpreting AST Structures

Interpretation is the process of determining the meaning of a Rosetta specification. To support interpretation, the Rosetta reflection system provides functions that reduce AST structures and map AST structures to mathematical structures called *co-algebras* (Table 13.1). Three functions are defined that perform basic evaluation. The `value` function reduces an AST structure to its simplest equivalent AST structure, or *normal form*. Similarly, the `typeof` function reduces an AST structure to the simplest AST structure representing its type. Finally, the `denotes` function reduces an AST structure to its normal form and attempts to denote that normal form in Rosetta's co-algebraic semantics.

Table 13.1 Interpretation functions associated with all Rosetta constructs

Function	Format	Meaning
Type	`typeof(x,c)`	*References the type of item x in the context of c*
AST Value	`value(x,e)`	*References the AST value of item x in context of environment e*
Semantic Value	`denotes(x,e)`	*References the semantic value of item x in the context of environment e*
String to AST	`parse(x)`	*Transforms item x into its associated AST structure*
AST to String	`string(x)`	*Transforms item x into a string representation*

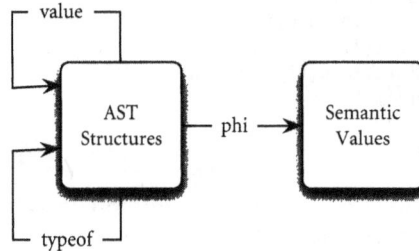

Figure 13.1 Relationships between evaluation functions, AST structures, and semantic values.

Figure 13.1 shows the relationship between the `value`, `typeof`, and `denotes` functions. The `value` and `typeof` functions reduce AST structures and the `denotes` function maps them to co-algebraic values.

13.2.1 The `value` Function

The objective of the `value` function is to find the *AST* representation of an AST structure's value given an environment *e*. It defines the transformation of a syntactic element into another, reduced syntactic element. For example:

```
value(<< expression 1+2 >>,e) == << expression 3 >>
value(<< expression inc(2) >>,e) == << expression 3 >>
value(<< expression 1+x >>,e) == << expression 1+x >>
```

The AST representing adding 1+2 is the AST for 3. The AST representing the value of 3 is itself, because 3 cannot be reduced. The AST structure representing the value of x+1 is also itself, unless the environment provides value for x.

An AST structure is in *normal form* when `value` does not reduce it further. In the previous examples, 1+2 is not in normal form because it can be reduced to 3; 3 is in normal form because it represents a literal value; and 1+x is in normal form because more information is required, specifically a value for x, before it can be reduced further.

An *AST value* is an AST structure in normal form that represents a Rosetta value. In the previous examples, 3 is a value because 3 cannot be reduced, regardless of information provided by the environment. Conversely, 1+x is not a value because 1+x can be reduced if the value of x is known. Like most languages, the set of values for Rosetta is identified prescriptively as a part of the language specification. In contrast, most languages define normal forms that are not values as errors representing an incomplete evaluation. As a specification language, Rosetta does not. Such normal forms are treated as descriptors of resulting value sets and reveal significant information about a specification.

13.2.2 The typeof Function

The typeof function accepts an AST structure and context and determines the AST structure's type. In many cases, typeof simply looks up declared types and calculates new types for expressions. However, the introduction of universally quantified parameters in function and facet signatures requires typeof to perform type inferencing. For example:

```
typeof(<< expression 1+2 >>,c) == << expression posint >>
typeof(<< expression 1+2.1 >>,c) == << expression real >>
typeof(<< expression inc(2) >>,c) == << expression natural >>
typeof(<< expression inc >>,c) == << expression <* (x::natural)::natural *> >>
```

Rosetta guarantees preservation of type with respect to value. Specifically:

$$typeof(x,c) == typeof(value(x,e),c)$$

where x is an expression, e is the evaluation environment and c is the context of x. The result of evaluating an expression has the same type as the original expression. This leads to a single axiom that expresses type correctness by asserting that any item's value must be contained in its type. Specifically:

forall(x::expression | typeof(value(x,e),c) =< typeof(x,c))

For declared items with constant types and values, the following two axioms link the AST defining the declared type and declared value to the type and value, respectively:

forall(x::expression | typeof(x) == (value(declared_type(x))))

forall(x::expression | value(x,e) == (value(declared_value(x,e))))

In both cases, finding the value of the AST structure defining the type or value provides the actual type or value.

Unlike a traditional evaluation function, value need not reduce to a value for all expressions. As a result, Rosetta cannot guarantee evaluation progress. This is to be expected, as few programming languages guarantee progress. Furthermore, specification languages are by definition not fully executable.

13.2.3 The denotes Function

In Rosetta, semantic values are co-algebra constructs used to provide definitions for AST structures. Rosetta uses a denotational approach for defining semantics, mapping each AST structure with its co-algebraic semantics. This is achieved using the the denotes function. Specifically, denotes transforms an AST structure representing a value into its co-algebraic meaning. Because type values are ordinary values, the AST structure may represent either a type or value. Thus, denotes(value(t,c),e) and denotes(type of(t,c),e) both map an AST structure to a structure in the co-algebraic domain.

All AST values map to semantic values. However, not all AST structures map to semantic values, because not all AST structures reduce to AST values. When an AST structure reduces to a normal form that is not a value, it can be thought of as representing a collection of AST values and thus semantic values that satisfy its properties. Although this is a problem for programming languages, it is necessary for specification languages, where a class of systems may satisfy a set of requirements.

Some simple examples of using the denotes function include:

```
denotes(<< expression 2 >>,e) == 2
denotes(<< expression 2e2 >>,e) == 200
denotes(<< expression [1,2,3] >>,e) == [1,2,3]
denotes(<< expression z+1 >>,e) == denotes(__+__(X),e)(denotes(z(X),e),1)
```

where X is the co-algebraic abstract state and $a(X)$ is the value of observe a in that state.

AST values and semantic values differ because of the nature of specification languages. The value function must be able to reduce a structure to something that is not a value. Thus, its signature must be a mapping from AST structure to AST structure. AST values must be of type AST structure to properly define value. However, as syntactic structures, they are still without semantic meaning. Semantic values are mathematical structures that provide meaning for AST values. This is why evaluating an AST value results in a semantic value.

13.2.4 The parse and string Functions

There are two additional functions defined for AST structures that merit some discussion. The parse function maps a string representation to its Rosetta abstract syntax representation. This construct is used to define how template expressions translate into their respected AST structures. The result of value (parse(x),e) is the Rosetta equivalent of evaluation. Specifically, it will parse a Rosetta expression to an AST structure and find its associated AST value.

The string function takes an AST structure and returns its associated text representation. This function is the inverse of parse and provides pretty-printing function for Rosetta expressions. The string function is rarely used in specifications and is included primarily for verification. The following example defines the correctness of the parse and string functions:

```
forall(x::expression | parse(string(parse(x))) == parse(x))
```

Table 13.2 Accessor functions associated with facet definitions

Function	Format	Meaning
Facet constructor	`make_complete_facet_declaration`	*Construct a facet AST*
Facet signature	`signature(f)`	*Signature of facet f*
Facet declarations	`declarations(f)`	*Set of items declared in f*
Facet terms	`terms(f)`	*Set of terms declared in f*
Export list	`exports(f)`	*Set of labels exported by f*
Parameters	`parameters(signature(f))`	*Set of f's parameters*
Domain	`facet_domain(signature(f))`	*Domain of facet f*

EXAMPLE 13.2

Rosetta Reflection
Functions for Manipulating
Facet Structures

Functions for manipulating facets provide an excellent example of writing and using reflective functions. A facet is simply an item of type facet whose value is a facet value. Like any Rosetta structure, functions associated with data types are used to access elements of a facet definition. These functions are described in Table 13.2 and are defined in `rosetta.lang.reflect`. The `signature`, `declarations`, `terms`, and `exports` functions access their respected AST structures in the facet declaration. To access parameter and type AST structures, `parameters` and `facet_domain` are called on the facet's signature.

Using these AST functions, we can make assertions about facet semantics, define functions over facet constructs and define the semantics of analysis techniques. First, define a facet's type as its domain. This following constraint asserts that for all facets the type of a facet is its domain:

```
forall(f::facet | typeof(f) = value(facet_domain(signature(f))))
```

Next, define a function to extend a facet's parameter list by adding a single new parameter to the end of the original list. The function simply appends a new parameter to the end of the sequence representing existing parameters. The function uses a universally quantified parameter to represent the type of the facet allowing the function to be applied to any facet of any type:

```
extend_facet_plist[d::domain](f::d,n::label,p::facet_parameter) :: d is
  << facet 'n'
    << facet_signature 'facet(parameters)^&[p]' 'facet_domain(f)'>>
    'exports(f)' 'declarations(f)' 'terms(f)' >>;
```

Finally, add a new form of static evaluation that checks to see if a declared value or parameter is constrained in a future state by some term in a facet:

```
potential_mod(f::complete_facet_declaration)::set(declaration) is
  filter(d::~(~declarations(f)) | ticked_or_dereferenced(d,terms(f))) +
  filter(d::~(~signature(f)) | ticked_or_dereferenced(d,terms(f)));
```

The `ticked_or_dereferenced` function determines if a declared item, d, appears in the context of a tick or dereference function in the terms of f. The `filter` application generates a set of declarations that satisfy the constraint. ∎

13.3 Domain Declarations

Defining new abstract model-of-computation and unit-of-semantics domains is not for the weak of heart. Fortunately, neither extension of the Rosetta base domain semi-lattice is often required. Defining new engineering domains is a common specification task that can easily be mastered. Writing new domains to express common abstractions in an engineering design domain involves adding declarations for those new abstractions. Defining new domains to serve as shared declarations between two specifications is similarly done by adding new declarations to existing domains.

Rosetta provides a mechanism within the base language for defining new domains. The syntax for a domain declaration is defined as:

```
⟦ use P ; ⟧*
domain D ⟦ parameters ⟧ :: T ⟦ with F⟧ is
    ⟦ export exports ⟦ with facetExports ⟧⟧
    ⟦ declarations ⟦ with facetDeclarations ⟧⟧
begin
    ⟦ terms ⟦ facetTerms ⟧⟧
end domain D ;
```

In this syntactic structure, D is the label naming the new domain, *parameters* is an optional collection of design parameters, *exports* list the exported names, *declarations* are items defined in the domain, T is the domain type, and *terms* are terms that define new axioms within the domain. T is the type of the domain being defined and the supertype of all domains and facets that extend D. As with facet definition, *exports*, *declarations*, and *terms* are all optional.

Throughout the domain definition are a number of optional **with** clauses that define how a facet extends a domain. The **with** clauses define how elements of the extending facet are elaborated to elements in the elaboration result. The **with** clause following the facet signature provides a name, F, used for the extending facet. This is necessary because the domain cannot know statically what facet is being extended. The remaining **with** clauses use elements from F to define new elements in the elaboration result.

The **with** clause associated with the **export** clause defines how to generate additional exported symbols from F. Typically, this **with** clause will simply specify the names exported from the extending facet. For example, if the domain exports p and F exports power, the result of elaboration will export p and power. If the **with** clause is not present, the default is to append the exports from D and the exports from F.

The **with** clauses associated with declarations and terms specify how to elaborate declarations and terms from F in the elaborated facet. This is where definitions for ticked expressions x', x@s, and x'dot are elaborated. The **where** clause specifies a function that is applied to F, accesses elements of F, and generates new clauses for the elaboration result. For example, the functions for transforming x' and x@s have the following form:

```
transform_tick(tick_expr::tick_operation)::tick_operator is
    << tick_operator 'operand(tick_expr)' 'at(next(s)) >>;

transform_tick_at(at_expr::infix_binary_operation)::tick_operator is
    << tick_operator 'left_operand(at_expr)' 'at
        ('right_operand(at_expr)' ) >>;
```

Here, transform_tick and transform_tick_at define the transformation of expressions of the form x' and x@s into expressions that use the at ticked expression. Both functions examine an argument expression, extract operands, and form an appropriate use of at. In the first case, x' is transformed to x'at(next(s)) by appending the next state function ticked expression. In the second case, x@s is transformed to x'at(s) by pulling the operation apart and assembling the operands into a ticked expression.

The function transforming the at ticked expression has the following form:

```
transform_at(at_tick::tick_operation)::expression is
    << function_application
        denotes('operand(at_expr)',
            'tick_argument_list((operators(at_expr))(0))(0)') >>
```

Transform_at defines the transformation of x'at(s) into denotes(x,s) where denotes defines the dereferencing operation; function_application is the abstract syntax structure being created. The denotes function is a special Rosetta function that provides link from specifications to the underlying co-algebraic semantics. In effect, denotes defines how a fully expanded symbol in a Rosetta specification is mapped to an observer over the co-algebra. By defining denotes, any domain can specify how a symbol is dereferenced with respect to its environment. Understanding the underlying mathematical semantics is not critical here.

If the **with** clauses are not present for declarations, then declarations from F are appended to those provided by the domain. Similarly, terms from F are appended if the **with** clause is not present. For the vast majority of domains written by users, **with** clauses are not required. Most engineering domains and even model-of-computation domains do not need to expand ticked expressions beyond what is provided by their associated unit of semantic domains. However, the definition mechanism is there, should it be needed.

The semantics of domain definition elements is identical to that of their analogous facet definition elements, with a few important exceptions. The **with** clauses that occur in domains are never present in facet definitions. All domain parameters are treated as kind **design** and cannot represent system inputs and outputs. Because of this, there are no kind annotations for domain parameters. The new domain's type, T, must be another domain. Because facets cannot serve as facet types, this restriction is easily enforced. Finally, introducing the new domain cannot introduce cycles in the domain semi-lattice. Each domain extends precisely one domain and cannot be its own supertype.

13.4 Defining Engineering Domains

Defining a new domain can be as simple as adding declarations to an existing domain. Many engineering domains are defined in precisely this fashion. In the `digital` domain from Example 13.3, the original domain is extended by adding new declarations and axioms over those declarations. Items declared in the original domain are not refined or constrained further. When such refinements are made, the new domain is considered a new model of computation domain.

EXAMPLE 13.3

Engineering Domain for Digital Design Defined by Extending the state_based Domain

One excellent example of an engineering domain is the definition of a `digital` domain for digital design extending the `state_based` domain. The `digital` domain defines a collection of utility functions that would be useful for defining RTL operations and integrating RTL components. The example here shows rising and falling clock predicates as well as a sampling of word-oriented operations.

```
domain digital()::state_based is
  rising(x::bit)::boolean is (x=1) and event(x);
  falling(x::bit)::boolean is (x=0) and event(x);
  wordAnd[1::natural](x,y::word(1))::word(1) is
    zip(__and__,x,y);
  wordOr[1::natural](x,y::word(1))::word(1) is
    zip(__or__,x,y);
  wordNot[1::natural](x::word(1))::word(1) is
    map(not,x);
  ...
begin
end domain digital;
```

None of the extensions to `state_based` reference its basic definitions directly. Each declaration defines a new function that references only parameters defined for that function. Thus, the extension is conservative and we can be certain that the `digital` domain as written is consistent.

The new `digital` domain is used to define facet models, like any other domain. Here, we use the `digital` domain to define a new component that accumulates its inputs using a conjunction on each rising clock edge. Clock and reset signals are used to synchronize and initialize the circuit, respectively. The `digital` functions `rising` and `wordAnd` are used to define the behavior of this component:

```
facet andAccum[1::natural](x::word(8); clk::bit; rst::bit)::digital is
  state :: word(8);
begin
  state' = if rst then x"FF" else
             if rising(clk) then wordAnd(state,x) else state end if;
           end if;
  o = state;
end facet andGate;
```

13.5 Defining New Model-of-Computation Domains

Defining new model-of-computation domains is not substantially more difficult than defining engineering domains. The difference is that items declared in unit-of-semantics domains are refined by model-of-computation domains. New items may also be added, but if they represent the only additions, the new domain is considered an engineering domain. If new basic computational elements are defined, then the new domain may be considered a new unit-of-semantics domain. This will rarely occur in typical specification situations.

EXAMPLE 13.4

The finite_state Model-of-Computation Domain Formed by Extending the state_based Unit-of-Semantics Domain

Examples of defining new model-of-computation domains occur repeatedly in the early sections of this chapter. One such example is the definition of finite_state from state_based. The only extensions performed in this domain declaration constrain state. First, state is constrained to the value instantiating the state_type parameter. Additionally, a constraint that the cardinality of state_type must be a natural forces the number of states to be finite.

No additional declarations are added to the original domain. Thus, this cannot be an engineering domain or a unit-of-semantics domain. Both such domains require addition of new declarations. The new domain simply places constraints on the state type, making it a model-of-computation domain.

```
domain finite_state(state_type :: type)::state_based is
  state :: type is state_type;
begin
  #state_type in natural;
end domain finite_state;
```

13.6 Defining New Unit-of-Semantics Domains

Writing new unit-of-semantics domains is achieved by starting with the null or static domain and adding declarations that define points of reference for observation. Specifically, unit-of-semantics domains define some type representing observation points, an operator that dereferences a label with respect to an observation point, and a function for changing the observation point. The state_based and signal_based domains provided by the base Rosetta system define state or event, @, and next to serve these roles.

Because @ and next are exceptionally common, built-in unit-of-semantics definitions, Rosetta provides special syntax for making their use concise. They are actually defined using the at ticked expression defining dereferencing. Specifically, these special syntactic forms have the following elaborations:

$$x@s \stackrel{\text{def}}{=} x'at(s)$$

$$x@next(s) \stackrel{\text{def}}{=} x'at(next(s))$$

Recall that a ticked expression is a special form that is defined by providing an elaboration rule in its associated domain. For example, x'at(s) is transformed into denotes(x,s) during elaboration prior to evaluation.

If a new unit-of-semantics domains can be specified by providing appropriate definitions for at and next, then the syntactic forms provided by Rosetta are immediately reusable in the new domain. Of course, a new domain can be defined from scratch using its own operators. In such cases, the new unit-of-semantics definitions must be used directly without additional syntactic sugar.

New unit-of-semantic definitions should be written only on the rare occasion when a fundamentally new vocabulary for describing computation is required. Remember that unit-of-semantics domains provide only vocabulary and must be accompanied by significant infrastructure to be useful. Most Rosetta users will never write new model-of-computation domains, much less new unit-of-semantics domains.

EXAMPLE 13.5

Defining the state_based Unit-of-Semantics Domain with Elaboration Functions

Among the most fundamental domains is the state_based unit-of-semantics domain. It defines the concept of state, the next state, and dereferencing symbols in a state. The following domain provides definitions for each of these quantities and defines the transformation needed to extend the domain with a facet definition:

```
domain state_based()::static with F is

    transform_tick(tick_expr::expression)::expression is
      << tick_operation 'left_operand(tick_expr)' 'at(next(s)) >>;

    transform_tick_at(at_expr::expression)::expression is
      << tick_operation
            'left_operand(at_expr)' 'at( 'right_operand(at_expr)' ) >>;
    transform_at(at_tick::expression)::expression is
      << function_instance
            denotes('left_operand(at_expr)','head(right_operand(at_expr))')
            >>;
    transform_deref(expr::expression)::expression is
      if is_tick_operator(expr) then
        if is_tick_operators(expr) == []
          then transform_tick(expr)
          else transform_tick_at(expr)
        end if
      elseif (is_infix_binary_operation(expr) and
                operator(expr) == commercial_at_token)
            then transform_at(expr)
        else expr
      end if;

    state_type :: type;
    next(state_type)::state_type;
    s :: state_type;
      with image(transform_deref,declarations(F));
  begin
    with image(transform_deref,terms(F));
  end domain state_based;
```

13.7 Defining Ticked and Dereferencing Expressions

A ticked expression is a special Rosetta form used to specify an item to be elaborated by a domain. Elaboration differs from evaluation in that it is a replacement of one form for another, prior to evaluation. What ticked expressions allow is the creation and manipulation of abstract syntax before evaluation. In effect, they manipulate Rosetta specifications as data using AST creation and access routines. All ticked expressions have one of the following forms:

$e't[\![(params)]\!]$

where e is an expression and $t[(params)]$ is a label followed by optional parameters.

During elaboration, the domain providing context for a ticked expression controls its elaboration. The ticked expression syntax is literally replaced with a syntax generated from it. The process is similar to macro expansion in assembly language — the expression is not evaluated, but rather is simply replaced by a newly generated expression prior to evaluation. A ticked expression does not represent a function call, but rather a new syntax element whose definition depends on its associated domain.

One example where ticked expressions are used extensively is when defining calculus equations. The time derivative of a variable is specified as:

```
x'dot
```

During elaboration, x'dot is literally replaced by:

```
x'deriv(t)
```

representing the time derivative of x. In turn, x'deriv(t) is defined using limit and can be elaborated to:

```
x'lim((x'at(t+delta)-x'at(t))/delta,delta,0)
```

The key characteristic of this process is that x is never evaluated. Thus, when we say something like:

```
x'dot = 1 / x
```

We are defining a differential equation rather than defining the application of a function. The distinction is subtle, but important.

13.7.1 Calculus Functions

A special class of ticked expressions for defining limits, derivatives, and integrals are provided for use with real valued functions. These expressions exist primarily to allow specification of ordinary and partial differential equations over real valued variables. Table 13.3 describes built-in Rosetta expression for finding limits, derivatives, and integrals with respect to functions and real variables. It is

Table 13.3 Calculus operations over functions

Operation	Syntax	Meaning
Limit	f'lim(x,n)	*Limit of f as x approaches n*
Derivative	f'deriv(x)	*First derivative of f with respect to x*
Time Derivative	f'dot	*First derivative of f with respect to time*
Antiderivative	f'antideriv(x)	*Antiderivative of f with respect to x*
Time Antiderivative	f'antidot	*Antiderivative of f with respect to time*
Integral	f'integ(x,l,u)	*The integral of f with respect to x ranging from l to u*
Time Integral	f'tinteg(l,u)	*Time integral of f with respect to t ranging from u to l*

important to note that Rosetta provides no standard mechanism for solving such equations, only a mechanism for specification. Individual Rosetta tool sets must provide solvers when analyzing specifications.

The limit of an expression is specified using the notation f'lim(x,c), where f is an expression over x and c is the limit. Specifically, f'lim(x,c) is interpreted as:

$$f'lim(x,c) \stackrel{\text{def}}{=} \lim_{x \to c} f(x)$$

The derivative of an expression is defined using limit in the canonical fashion. The following definition is provided for all real valued expressions and real valued non-zero delta:

$$f'deriv(x) \stackrel{\text{def}}{=} f'lim((x+delta)-f(x))/delta),delta$$

In the derivative expression, f is the expression and x is the parameter subject to the derivative. For the above definition, the following holds:

$$f'deriv(x) \stackrel{\text{def}}{=} \frac{d}{dx} f(x)$$

The derivative expression is generalizable to partial derivatives. Assuming that g is defined over multiple parameters, such as g(x::real;y::real;z::real):: real, then:

$$g'deriv(x) = \frac{\delta}{\delta x} g(x,y,z)$$

Antiderivative, or indefinite integral, is the inverse of derivative. The antiderivative of f with respect to x is expressed as:

$$f'antideriv(x) \stackrel{\text{def}}{=} \int f(x)dx$$

f being the target function, x being the variable integrated over, and c being the constant of integration.

An antiderivative is the dual of derivative. The following definition is provided for all real valued functions:

```
(f'deriv(x))'antideriv(x) == f
```

The definite integral of f with respect to x over the range l to u is expressed as:

$$\texttt{f'integ(x,l,u)} \overset{\text{def}}{=} \int_l^u f(x)dx$$

The definite integral is defined as the difference of the indefinite integral applied at the upper and lower bounds:

$$\texttt{f'integ(x,l,u)} \overset{\text{def}}{=} \texttt{(f'antideriv(x))(u) - (f'antideriv(x))(l)}$$

It is possible to express a definite integral over an infinite range using the notation:

$$\texttt{f'integ(x,infinity,-infinity)} \overset{\text{def}}{=} \int_{-\infty}^{\infty} f(x)dx$$

It should be noted that limit, derivative, antiderivative, and integral expressions are defined over real valued functions and values in domains with continuous temporal references. Further, the functions provide a mechanism for expressing these operations and some semantic basis for them. Solution mechanisms are not provided as a part of the language definition.

EXAMPLE 13.6

Defining Simple Calculus Equations

Performing higher-order derivatives is a matter of simply applying the derivative multiple times. For example:

```
f'deriv(x)'deriv(x)
```

is the second derivative of f with respect to x.

Differential equations are defined by creating terms that equate derivatives with expressions:

```
f'deriv(x) = 1/x
```

defines the single differential equation:

$$\frac{df(x)}{dx} = \frac{1}{x}$$

The classical calculus expression for calculating the x position of an object at time t starting from position 0 is:

```
x(t) = (t^2*x'deriv(t)'deriv(t))/2 + t*x'deriv(t) + 0
```

■

EXAMPLE 13.7

Defining Time Derivatives and Integrals

To simplify definitions, Rosetta provides a collection of calculus expressions that use the temporal reference by default on variables. In the same manner that x represents the first derivative of x with respect to t, x'dot represents x with respect to the current domain's temporal reference. In continuous time, this is t. In state_base, this is s.

```
x = (t^2*(x'dot'dot))/2 + t*x'dot + 0
```

Similarly, the antiderivative with respect to time is x'antideriv and the definite integral is x'tinteg(1,u).   ∎

13.7.2 **Dereferencing**

Name dereferencing is an exceptionally common Rosetta definition activity that is entirely domain specific. For this reason, we elaborate dereferencing expressions to ticked expressions that can be defined at the domain level. There are two forms used for explicit name dereferencing. The notation x' refers to an item in the next state, and x@s refers to an item in some arbitrary state s.Thus, x' is equivalent to x@next(s). In these expressions, the dereferencing operator @, the next state function next, and the state s are domain specific and cannot be defined across the entire Rosetta language. To define their meanings, Rosetta elaborates the dereferencing expression x@s to x'at(s). Rosetta defines the meaning of x@next(s) and x@s using a ticked expression elaboration in state_based domain. Specifically, the definitions for dereferencing expressions are:

$$e@s \overset{\text{def}}{=} e\text{'at}(s)$$
$$e\text{'} \overset{\text{def}}{=} e\text{'at}(\text{next}(s))$$
$$e\text{'at}(s) \overset{\text{def}}{=} \textit{denotes(e,s)}$$

The at expression provides the actual definition of dereferencing and is declared in several unit-of-semantics domains. Thus, by specializing the definition for at in a domain, dereferencing is defined on that domain. (The function *denotes* represents the dereferencing operation, but its specific definition is beyond the scope of this text.) For example, at and next are declared in the state_based domain and given abstract definitions. By specializing at, subsequent domains such as finite_state or discrete_time can define specific dereferencing operations by providing a new definition for at. Specializing next allows a domain to specify properties for state change resulting in different models of computation. However, the syntax presented to the user remains the same.

Outside the context of the state_based domain and domains extending it, the notation x' has no meaning unless at and next are declared and defined. For example, in the static domain, x' and x@s have no meaning because at and next are not declared in that domain. However, in a signal_based or spatial domain that defines a fundamentally different semantics, at and next can be defined in a completely different manner. For example, in a spatial domain, the state type could be a Cartesian coordinate, at would dereference with respect to the coordinate, and next would have no definition.

13.8 Consistent Domain Extension

Ensuring consistency in newly defined domains and facets is a constant issue when defining specifications, because checking consistency is exceptionally difficult. Some steps can be taken to help ensure consistency. One in particular is following some simple rules to ensure that the extensions provided are conservative. If these techniques are followed and the original domain is consistent, the resulting domain will also be consistent.

The simplest way to assure a conservative extension is to avoid using existing items in new declarations. If the original domain is consistent and none of the original items is referenced in new declarations, then no new properties can be asserted over existing items. Thus, no properties making the original specification inconsistent can be added. Although this restriction may seem overly conservative, engineering domains, such as the digital example where new definitions include only functions, are easily defined. As most user-defined extensions are for engineering domains, a conservative extension is frequently possible.

If it is simply not possible to avoid referencing declarations from the original domain, the extension can still be conservative. If new definitions do not redefine or add to existing definitions, the extension will be conservative. Consider the definition of a domain that defines a Stack type by defining the empty item and push, pop, and next operations:

```
domain hasStack(elem::type)::static is
  Stack :: type;
  push :: <*(e::elem; s::Stack)::Stack*>
  next :: <*(s::Stack)::elem*>
  pop :: <*(s::Stack)::Stack*>
begin
  pop(push(e,s)) == s;
  next(push(e,s)) == e;
end domain hasStack;
```

The easiest way to define the behavior of pop is to provide a definition of its behavior over push. This is done in the two terms defined in hasStack.

Now define a new domain that extends our original domain to include a size operation. This cannot be defined without referencing push. However, it need not contradict existing properties if it avoids the use of other existing functions. Specifically, the definition of size is added to the original domain in the following manner:

```
domain hasStackSize(elem::type)::hasStack(elem) is
  size :: <*(s::Stack)::natural*>
begin
  ax1: size(empty) == 0;
  ax2: size(push(e,s)) == size(s)+1;
end domain hasStackSize;
```

Although ax2 references the pre-existing operation push, it does not reference pop or next. Because the terms from the original specification only define pop and next directly, it is impossible to creat a contradiction if we do not constrain pop and next further.

Ultimately, caution must be taken when extending domains. Although facet definitions have similar issues, when a facet is inconsistent, one model is inconsistent. When a domain is inconsistent, every facet and domain of that type and its subtypes is inconsistent. One should not avoid writing new domains, but simply approach the process carefully.

The Facet Algebra

The *facet algebra* is a collection of operations for composing facets from one or more domains. The facet *product* and *sum* operations define classical conjunctive and disjunctive composition operations. Taking the product of two facets defines a new facet that satisfies both of the original specifications. Similarly, taking the sum of two facets defines a new facet that satisfies either of the original specifications. Using facet product and sum provides direct language support for concurrent engineering where multiple specifications must be simultaneously valid.

The **if**, **case** and **let** forms define the mechanisms for selecting between models and defining local models respectively. Facet-typed applications of **if** and **case** allow selection of behavior based on boolean conditions. Facet typed applications of **let** allow definition of local symbols over facet-typed expressions. The only distinction between facet algebra instances and traditional instances of these constructs is that facet algebra instances will have facet types.

Unlike product and sum, *Homomorphism* and *isomorphism* do not form new facets, but define relationships between facets. A homomorphism exists between two facets when the properties of one are implied by the other. Homomorphisms in the domain semi-lattice are critical for moving information between domains. An isomorphism exists between two facets when a homomorphism exists both ways. When facets are isomorphic, they are considered equivalent.

14.1 Facet Products and Sums

Products and sums originate from category theory and are widely accepted mechanisms for specification composition. It is not necessary to understand these concepts deeply to understand their utility in Rosetta. It is sufficient to understand initially that products and sums combine two specifications into a single specification that comprises both original specifications. The product defines a new specification that is simultaneously both of the original specifications, while the sum defines a new specification that is either of the orignal specifications.

An excellent example of a product from programming languages is a record, class, or tuple structure. A tuple type is literally the Cartesian product of its constituent types. The **tuple**(1,''a'',2.1) is an element of the type:

```
integer × string × real
```

The best way to distinguish a product is to note that a tuple is not a tuple without one of its elements. In other words, (1,''a''), (1), and (''a'',2.1) are not elements of the product type. All three values must be present.

In contrast, a sum composes specifications into a single specification that is either of the original specifications. An excellent example of a sum from programming languages is a union type or a constructed type. A union is the tagged union of the original specifications. The sum type:

```
integer + string + real
```

contains values from the original three types. To assure value distinctness, programming languages will tag values to indicate the source type of the original value.Thus, (left 1), (middle ''a''), and (right 2.1) are all members of the sum type. In this case the tag indicates the source type of the value by indicating the position of the type in the sum type declaration.

Syntax for facet product and sum operations specifies two facets, the operator and an optional **sharing** list:

```
F * G ⟦ sharing { symbols } ⟧
F + G ⟦ sharing { symbols } ⟧
```

The facet product, $F * G$, states that both specifications F and G simultaneously define a system and must simultaneously hold. The facet sum, $F + G$, states that both specifications F and G separately define a system. The optional **sharing** keyword defines the vocabulary that both specifications share. In both cases, the **sharing** clause identifies items that are shared among the specifications and must satisfy properties from both specifications.

14.1.1 The Shared Domain

Any two facets defined from domains in the same domain semi-lattice must share at least one common domain. Even if that domain is static or null, the common domain will always exist. Graphically, the common domains are easily found by tracing the lattice structure upward from the domains of two facets, until a common domain is discovered. This relationship is defined mathematically by identifying a partial ordering relationship over the domains in the semi-lattice and the existence of a minimum element resulting in a join semi-lattice.

When a domain is defined, a superdomain must be identified as the supertype of the new domain. In this specification fragment:

domain $D_{sub}(...) :: D_{super}$ **is**
    ...

D_{sub} is a new domain, with D_{super} as the domain it inherits from; D_{sub} is a subtype of D_{super}.

When one domain extends another in this manner, a partial ordering relationship exists. Specifically, the partial ordering is defined such that when $D_{sub} :: D_{super}$, the ordering relationship $D_{super} \sqsubseteq D_{sub}$ holds. Because all domains extend null, null $\sqsubseteq D$ is true for all domains and $D \sqsubseteq$ null is never true for any domain. Thus, null defines the minimum domain, or bottom of the semi-lattice.

Using $\sqsubseteq$ we define the meet, or *greatest common supertype* $(D_1 \sqcap D_2)$, of domains D_1 and D_2 as an element of the set:

$$\{D \mid D \sqsubseteq D_1 \wedge D \sqsubseteq D_2\}$$

such that every other member of the set satisfies:

$$D \sqsubseteq (D_1 \sqcap D_2)$$

Thus, $D_1 \sqcap D_2$ must be the maximum value on the set with respect to the $\sqsubseteq$ partial order. Because all domains extend null, $D_1 \sqcap D_2$ exists for any two domains in the semi-lattice, even when $D_1 \sqcap D_2$ is null.

We care about the greatest common supertype because it defines the maximal set of specification objects shared by two domains. This *shared specification* defines the common items that both specifications observe. If two facets are written that extend the discrete_time domain, discrete_time becomes the greatest shared supertype and they both share the same definition of time. Most importantly, it is *the same* definition of time. Thus, any property asserted over the time value in one facet must be consistent with the other under composition. Any item that appears in the greatest common supertype is shared in the same manner. In this way, specifications under composition can share the same type.

If two specifications do not share the same type, then by definition they share their greatest common supertype. If a specification written in the state_based domain is composed with a specification in the infinite_state domain, the shared items are state and next. The infinite, discrete nature of state is not known to the specification written in the state_based domain. As a result, the type of any composition of these two specifications must be state_based, losing more detailed information about the infinite_state specification.

A simple example demonstrates the role of the shared specification in facet composition. The following two facets define parity checkers that differ only in when state is updated. The first, parity1, updates its state on the rising clock edge, while the second, parity2, updates on the falling edge. As defined, there is nothing wrong with either facet.

EXAMPLE 14.1

Impact of Two Facet Specifications Sharing a Common Supertype

```
facet parity1 (clk::input bit; o::output boolean)::infinite_state (natural) is
  next(s::integer)::integer is s+1;

begin
  up: s' = if rising(clk)then next(s) else s end if;
  out: o' = odd(s);
end facet parity1;

facet parity2(clk::input bit; o::output boolean)::infinite_state (natural) is

  next(s::integer)::integer is s+1;
begin
  up: s' = if falling(clk) then next(s) else s end if;
  out: o' = odd(s);
end facet parity2;
```

It bears repeating that there is nothing wrong with either of these specifications in isolation. They simply provide two definitions for what parity checking is. What if we compose these facets using a product to specify a single system that must satisfy both requirements:

```
facet parity(clk::input bit; o::output boolean)::infinite_state(natural)is
  parity1(clk,o) + parity2(clk,o);
```

The parity facet composes the original facets using the product operation. The greatest shared supertype is:

```
infinite_state (natural)
```

All definitions including next, state, and s are shared among the two specifications. This implies that definitions over those items must be consistent across facet specifications. Simply changing the triggering event for state update causes an inconsistency in the composition. Specifically, the value of the current state will be out of sync in the two models due to differing update times. Discovering this inconsistency is exactly the desired result, as it allows us to see a system-level problem caused by a local specification decision. ∎

The distance between specifications in the semi-lattice is referred to informally as the *intellectual distance* between the specifications. As the name implies, the greater the intellectual distance between specifications, the more difficult it is to compose specifications. Although any two domains must share some greatest common superdomain, the intellectual distance between them may make their composition virtually useless. If two specifications share only the null or static domain, the intellectual distance is maximized. In essence, they share no common items other than basic mathematical definitions. Such problems can be avoided by defining functions that move specifications from domain to domain in the semi-lattice.

EXAMPLE 14.2

Facet Composition Identifying the Shared Domain of the Original Facets

The following table illustrates the calculation of shared specifications from several domains in the domain semi-lattice. In general, determining the greatest common supertype simply involves following the tree upward from the two domains and stopping at the first point where the two paths intersect.

Domain A	Domain B	Shared Specification
state_based	state_based	state_based
state_based	finite_state	state_based
finite_state	state_based	state_based
finite_state	infinite_state	state_based
finite_state	signal_based	static

Products and sums are calculated using the greatest common supertype of the composed facets. The product defines two facets that together define the composite specification. The facet definitions are linked by their shared parts. Specifically, both facets define properties over items from the greatest common supertype that must be satisfied under composition. Similarly, the sum defines two facets that define cases in the composite specification. Again, the facet definitions are linked by properties defined over their shared parts. Semantically, products and co-products are constructed using pullbacks and pushouts, respectively. Detailed knowledge of these constructions is not required to compose Rosetta specifications.

14.1.2 The Sharing Clause

In many cases, it is desirable for facets to share more than just those items defined in a standard domain. Types, functions, and elements of system state are all examples of items useful for sharing. There are two approaches to adding something to the shared specification for use in a product or sum. The first is to write a new, specialized domain that adds shared items to the domain involved in the product or sum. The second is to use a **sharing** clause to implicitly add items to the shared part. Both have their advantages; however, the **sharing** clause tends to be preferred due its general nature and simplicity of use.

EXAMPLE 14.3

Facet Composition Using Facet Product

Composing specifications using the product operation defines a new specification that satisfies both original specifications. The following facets define a simple component and a constraint that the component must satisfy. For this system to be correct, both specifications must be satisfied, indicating the use of a product for composition.

```
facet component(i::input integer; o::output integer)::state_based is
begin
end facet component;

facet constraint(v::design real)::static is
begin
end facet constraint;
```

The facet system defines the product of **component** and constraint. For system to be consistent, both **component** and constraint must be satisfied over the same shared specification. In this case, the shared part is static and consistency is trivial to assert as long as elements of static are not redefined.

```
facet system i::input integer; o::output integer)::static is
   component(i::input integer; o::output integer) * constraint(2.3);
```

The specification is much more interesting if a functor is used to generate a state_based specification from the constraint. The concretization function gamma takes an abstract specification and makes it more concrete. In this case, it takes the static constraint specification and asserts the constraint in every state. The result is a new system facet in the state_based domain. This new specification does not sacrifice details of the component specification to achieve composition.

```
facet system(i::input integer; o::output integer)::state_based is
   component(i::input integer; o::output integer)
   * gamma(constraint(2.3))::state_based;
```

The specifics of defining gamma and how it is used will be discussed in detail later in this chapter. ∎

The **sharing** clause is used with facet sum and product operations to extend the shared domain by introducing new items that are shared among the two facets involved in the composition operation. The **sharing** clause does not declare new items, but simply adds existing items to the shared specification from facets under composition. Specifically:

$F + G$ sharing S

forms the sum of facet's F and G, with S extending the greatest common domain of F and G. Elements of S must be defined in both F and G, but need not have a common type. The format of S is a set of names shared between the specifications. Thus:

$F + G$ **sharing** $\{H,I,J\}$

defines the sum F+G, where {H,I,J} name the new shared items in both specifications that are added to the shared specification. Similarly:

$F * G$ **sharing** $\{H,I,J\}$

defines the product F*G, where {H,I,J} names shared items in both specifications that do not exist in the greatest common supertype. In both cases an error exists if the shared items are not declared in both specifications. Items from the domain can be listed in the **sharing** clause. However, both F and G inherit those definitions from the shared domain without the **sharing** clause.

The semantic definition of the **sharing** clause pushes shared definitions from the facets being composed into the domain. For example, in the following specification, the product of test1 and test2 is formed **sharing** the state variable x. Thus, x refers to the same object in both specifications under the product.

Conversely, y is not shared, causing both specifications to maintain their local, independent definitions.

```
facet interface test1(a::input bit; b::output bit)::state_based(bit) is
   x::integer;
   y::real;
end facet interface test1;
facet interface test2(a::input bit; b::output bit)::state_based(bit)is
   x::integer;
   y::real;
end facet interface test2;
test3 :: state_based(bit) is test1*test2 sharing {x};
```

An alternative approach is to write a new domain to accomplish the same task by extending the greatest common supertype domain. Rather than use the **sharing** clause to push items into the domain implicitly, a new domain explicitly defines those items:

```
domain test_domain()::state_based(bit) is
 x::integer
begin
end domain test_domain;

facet interface test1(a::input bit; b::output bit)::test_domain is
 y::real;
end facet interface test1;

facet interface test2(a::input bit; b::output bit)::test_domain is
 y::real;
end facet interface test2;

test3 :: state_based(bit) is test1*test2;
```

The new domain test_domain defines a domain specific to this specification task where x is added to the state_based specification. The product no longer requires the **sharing** clause because x now appears in the domain and is shared by default. Although there is nothing wrong with this approach, using the **sharing** clause avoids writing specialized domains and defeating the idea of domain reuse. Semantically, the second definition is identical to the first.

EXAMPLE 14.4

Facet Composition Using
Facet Sum

In contrast to the product operation for facets, then sum operation combines specifications disjunctively. Specifically, the result of a sum is a new specification that exhibits properties of one or the other specification. In the following definitions, componentA and componentB define alternate component behaviors.

```
facet component A(i::inputinteger;o::output integer)::state_based is
begin end facet componentA;

facet componentB(i::input integer;o::output integer)::state_based is
begin end facet componentB;
```

The **component** facet, formed from the sum of facets componentA and componentB, can be observed as either of the original components.

```
facet component (i::input integer; o::output integer)::state_based is
   componentA (i::input integer; o::output integer)
   + componentB (i::inputinteger; o::output integer)
```

When a facet resulting from a sum appears in a specification, it may exhibit behaviors from either of its constituent facets. Thus, the environment of the specification must deal with both possibilities. The environment cannot assume that either behavior set will occur, and must be prepared for both. The facet sum does not allow refinement of behavior, but actually expands a definition by allowing multiple, correct behaviors.   ■

14.1.3 Implicit Sharing

Parameters appearing in facet interfaces are implicitly shared if they have the same name. This simply asserts that unique parameters have unique names. From a physical design perspective, this makes sense. For example, the clock signal should be the clock signal in every model where it is applicable, regardless of modeling domain.

For example, the following models define intersecting parameter lists. Under composition, either sum or product, parameters with the same name are treated as a part of the shared domain. There is an implicit **sharing** clause indicating the parameters across specifications.

```
facet interface devFunction
   (x::input integer; y::output integer)::state_based is
end facet interface devFunction;

facet interface devPower
   (x::input real; leakage,deltaP::design real)::continuous is
end facet interface devPower;

dev(x::input number; y::output number; leakage,deltaP::design number) ::
   state_based is devFunction(x,y)and devPower(x,leakage,deltaP);
```

The input parameter x appears in both specifications and is a part of the shared domain associated with the product. The output y only appears in the functional model and is thus not shared. Likewise, the design parameters leakage and delta do not appear in the functional model and are thus not shared. However, all parameters appear in the parameter list associated with dev.

The parameter list associated with the product specification, dev, uses the type number to constrain values in its input and output. Another type could be chosen, but the number type expresses only the desired requirement. Specifically, the values of connected parameters must be number, but need not specifically be integer or real. The parameters represent distinct observations of the product's abstract state. However, those observations must be mutually consistent with any translation functions that may be defined. For example, the integer and real values associated with x must satisfy the translator defined between the state_based and continuous domains. If no such translator is defined, then any assignment is legal, thus it is important to consider translators during specification.

The following Rosetta interface specifications define two composite specifications that will be interconnected. The first, fandc0, defines a model that is the product of f0 and c0. The models for this component, c0 and f0, implicitly share the parameters x and y. Because c0 is static and f0 is state_based, the interaction between parameters in the two models is governed by the active interaction between static and state_based.

```
facet interface c0(x::input integer; y::output integer)::static is
end facet interface c;

facet interface f0(x::input real;y::output::real)::state_based is
end facet interface f;

fandc0(x::integer; y::integer) :: static is f0(x,y) and c0(x,y);
```

The fandc1 facet is analogous to fandc0 except that the models are defined in the discrete_time and state_based domains. Again, the interaction between models is governed by the active interaction between state_based and discrete_time.

```
facet interface c1(x::input integer; y::output integer)::discrete_time is
end facet interface c1;

facet interface f1(x::input real; y::output::real)::state_based is
end facet interface f1;

fandc1(x::integer; y::integer) :: static is f1(x,y) and c1(x,y);
```

The interconnect facet simply connects the two previously defined components in sequence, with the output of fandc0 used as the input to fandc1.

```
facet interconnect(x::input integer;y::output integer):: static is
  z :: integer;
begin
  cmp0: fandc0(x,z);
  cmp1: fandc1(z,y);
end facet interconnect;
```

Because both fandc0 and fandc1 have models in the state_based domain, any constraints on parameters in that domain must be mutually consistent. However, fandc0 does not have a discrete_time model, and fandc1 does not have a static model. Because an interaction is defined between state_based and the other two domains, information is implicitly shared across the parameter. Specifically, all the assertions made by all models must be mutually consistent in the presence of the interactions. Because an interaction exists between state_based and static in fandc0 and between state_based and discrete_time in fandc1, an implicit transitive interaction exists between static and discrete_time.

Revisiting our activity-based power modeling example, we would like to add the capability of including internal switching in the power calculation. The model presented in Chapter 12 only monitored the output signal for value changes. The system facet is a structural model of a simple system that sequences multiplication and division operations. Because the internal signal changes values as well as the external signal, there will be an associated power drain.

```
facet system(i::input integer; o::output integer)::state_based is
  x::integer;
begin
  c1: mult(i,3,x);
  c2: div(x,4,o);
end facet system;
```

The new systemPower model defines a local variable, x, that parallels the local definition in system. However, this is a local variable and is not the same item that is defined in system.

```
facet systemPower[Ti,To::type](i::input Ti; o::output To;
                  leak,switch::static real)::state_based is
  export power::real;
  x::top;
  powerUpdate[T::type](o::T)::real is
    leakage + if event(o) then switch else 0 end if;
begin
  power' = power
        + powerUpdate(x,leak,switch)
        + powerUpdate(o,leak,switch);
end facet systemPower
```

To compose the specifications and make systemPower aware of changes in x from facet system, we use a **sharing** clause to push x into the shared part.

```
powerAware[Ti,To::type](i::input Ti;o::output To)::state_based is
  system(i,o) * systemPower(i,o) sharing {x};
```

Now x in powerAware and x in system refer to the same item. When system changes the value of x, that change will be observed by systemPower and power consumption updated as appropriate. ∎

14.2 Facet Homomorphism and Isomorphism

A *homomorphism* exists between two facets f_1 and f_2 when all the properties of f_2 are also properties of f_1. This is denoted using the facet homomorphism or facet implication relation:

$$f_1 \Rightarrow f_2$$
$$f_2 \Leftarrow f_1$$

An *isomorphism* exists between two facets f_1 and f_2 when they exhibit exactly the same properties or when f_1=>f_2 and f_1<=f_2 hold simultaneously. This is denoted using the facet isomorphism or facet equality relation:

$$f_1 \text{ <=> } f_2$$

Homomorphism and isom orphism are used primarily to express desired relationships between facets rather than to construct new facets. For example, these relationships can be used to express correctness conditions in the implications section of a component structure.

Facet homomorphism plays a critical role in the formation of the domain semilattice. Whenever one domain is extended to form another domain, a homomorphism exits between the new domain and the original domain. It follows that everything expressed in the original domain can be expressed in the new domain. When functors are defined between domains, the existance of a homomorphism assures that information from the source domain is not lost in the destination domain when facets are transformed. When an isomorphism exists between two domains, fact can be moved back and forth between them without losing information.

14.3 Conditional Expressions

The **if** expression can be used with facet values in the same manner as they are used with any other value. In the following equation, the `test` value determines the value of the **if** expression:

```
if t then F else G end if;
```

If t is **true**, the expression evaluates to F. Otherwise, the expression evaluates to G. Like any Rosetta expression, **if** can have only one type requiring F and G to share a common type. The type of the **if** expression is this shared type. If the type cannot be inferred, ascription can be used to assist in the analysis process. However, if a facet's type is abstracted away, it is gone and may not be reconstructible at a later point. Remember that ascription is not casting — if the assertion `f::D` cannot be verified, it is not considered correct.

EXAMPLE 14.6

Uses of Homomorphism and Isomorphism Relationships

Some interesting relationship can be defined using isomorphisms and homomorphisms. The simplest is that any facet is isomorphic to itself. The **forall** quantifier is used to state that the self-isomorphism relationship holds for all facets:

```
forall(f::static | f <=> f)
```

Similarly, a homomorphism exists between a facet and its domain. Because the facet extends the domain, we know that any property present in the domain is also present in the facet. The notation may seem odd because the domain is used as both a type and a facet. This is perfectly acceptable:

```
forall(f::D | f => D)
```

This relationship is important, as it defines the domain semi-lattice. Specifically, it serves as the partial order giving the semi-lattice its structure an defining the greatest element. Correctness conditions can be expressed using both homomorphisms and isomorphisms. Assume that R is a facet defining the requirements of a component and S is a structural facet defining an implementation architecture. The following relationship states that the architecture must exhibit all properties of the requirements specification:

```
S => R
```

Alternatively, we can use an isomorphism in conjunction with an abstraction functor to define a similar requirements relationship:

```
alpha(S) <=> R
```

Here the abstraction functor alpha removes detail from the S facet. This less detailed facet is then equated with the requirements facet. The idea is that extraneous properties that are not germane to correctness are removed by the abstraction function.   ∎

The **case** expression works identically to the **if** expression. The **case** argument selects one case and the associated facet expression is returned. Again, the facets in the **case** statement must share a common type and the **case** statement is of that type.

14.4 Let Expressions

The **let** expression defines local items over expressions or collections of terms. These local items may have any type including a facet type. Thus, the **let** expression may declare new facets for use in definitions. This is technically not a facet algebra operation, but is worth mentioning, as it can have utility in complex specification definition.

EXAMPLE 14.7

Conditional Specifications Using both **if** and **case** Expressions

The system facet below is a partial specification of a system where one of two system models may be selected using the parameter lowPower. The intent is to allow a user to select between low and normal power configurations of the same system. By instantiating lowPower with **true**, the specification includes a low-power CPU model. Alternatively, instantiating lowPower with **false** selects a normal, higher-power CPU.

```
facet system(lowPower::design boolean)::discretetime is
begin
  c₁: component1(...);
  cpu: if lowPower
          then lp_arm(...)
          else cots_arm(...)
       end if;
  c₂: component2(...);
  ...
  cₙ: componentN(...);
end facet system;
```

Such conditional specifications are also useful for selecting between different vendor models of a component. The following specification allows this by using a **case** statement to select a model based on a single input parameter.

```
package multiVendor()::static is
   vendor :: type is data
      Mot() :: Motp
      |TI() :: TIp
      |ibm() :: ibmp
   end data;
end package multiVendor;

use multiVendor();
facet system(cpuModel::design vendor)::discrete_time is
begin
  c₁: component1(...);
  cpu:case cpuModel of
         {Mot()} -> mot-arm(...)
         {TI()} -> ti-arm(...)
         {ibm()} -> ibm-arm(...)
      end case;
  c₂: component2(...);
  ...
  c_N: componentN(...);
  end facet system;
```

The multiVendor package defines a single type used to specify the CPU vendor in the specification. The **case** statement then uses the defined vendor values to select from three different cpu implementations. ∎

The facet adder2 defines a structural 2-bit adder using a half-adder to sum the first bit and a full adder to sum the second. The **let** clause is used to define a local half-adder by instantiating a full adder with a carry in of 0.

EXAMPLE 14.8

Local Facet Definition Using a **let** Expression to Define a Half-adder

```
facet adder2(x0,y0,x1,y1,z0,z1,z2)::static is
  c0 :: bit;
begin
  let halfAdder(x,y::input bit; z,cout::output bit)
        be adder(x,y,0,z,cout) in
     a0: halfAdder(x0,y0,z0,c0);
     a1: adder(x1,y1,c0,z1,z2);
  end let;
end facet adder2;
```

There are numerous mechanisms for writing this adder, including using two full adders rather than creating a half-adder. This mechanism is a perfectly reasonable alternative.   ∎

14.5 Higher-Order Facets

Higher-order facets are facets having facet type parameters, in the same spirit as higher-order functions having function type parameters. Such facets are exceptionally useful for defining high-level architectures and other common system structures.

An exceptionally common structural architecture is sequential ordering of two or more operations. We will refer to this as the *batch sequential* architecture defined by facet batchSequential.

```
facet batchSequential[T0,T1,T2::type]
                      (f1(x::T0; y::T1)::state_based;
                       f2(y::T1; z::T2)::state_based;
                       i::input T0; o::output T2)::state_based is
    x::T1;
  begin
    c1: f1(i,x);
    c2: f2(x,o);
  end facet batchSequential;
```

The batchSequential facet signature defines three universally quantified parameters, two facet types and an input and output. The universally quantified parameters, T0, T1, and T2 define the input, exchange, and output types associated with the facet. These types are not restricted in any way and can take any type value desired. Making them universally quantified allows discovery or direct specification of the desired type.

The two sequentially connected components are f1 and f2. Facet f1 takes component input, performs a transformation, and generates output for f2. Facet f2 takes output from f1 and generates component output. The universally quantified parameters must be instantiated to satisfy type conditions.

EXAMPLE 14.9

Using the Batch Sequential
Architecture

Now that we have a batch sequential architecture, we can put it to work by instantiating it in other systems or defining new components. The partial facet, system, uses sequenced negate facets to implement a delay buffer. The buffer component instantiates batchSequential with two negation operations and specifies inputs and outputs to the new device.

```
facet negate(i::input bit; o::output bit)::state_based is
begin
  o'=not(i);
end facet negate;

 facet system(...)::state_based is
begin
  ...
  buffer: batchSequential(negate,negate,in,out);
  ...
end facet system;
```

Alternatively, we define a buffer facet that can be reused throughout a system specification. Here a new facet, buffer, is defined by specifying its value instantiating batchSequential.

```
buffer(i::input bit,o::output bit)::state_based is
   batchSequential(negate,negate,i,o);
```

The new buffer facet is defined over two parameters that serve as input and output for the batchSequential instantiation. The component parameters to batchSequential are again instantiated with copies of negate.   ∎

Domain Interactions

The motivation for *domain interactions* is that decisions made locally in one domain can have impacts on system-level requirements and other domain-specific requirements. For example, heat dissipation is typically not thought of as a constraint that is impacted by the choice of a software algorithm. Unfortunately, software execution can dramatically impact heat dissipation through its interaction with the CPU it runs on, and indirectly through interfaces to other system components. Other examples include impacts of local power consumption on system-level power constraints, impacts of component function on system security, and impacts of electromagnetic interference among chips on a board. In each case, local decisions impact the overall system.

Rosetta defines an *interaction* as a collection of mechanisms for describing how information flows between modeling domains. Each interaction defines three kinds of information transfer. *Translators* define how information is transformed as it flows through a parameter shared by two facets. Translators handle both information translation and coordination of the communication process. *Functors* define transformations of facets from one domain to another. Because facets are simply Rosetta values, these transformations take the form of functions whose domain and range include facet types. *Algebra combinators*, or simply *combinators*, define mechanisms for combining two facets to produce a new facet. Combinators are like functors, except that the domain of a combinator must include two facets that will be composed into a single facet.

15.1 Projection Functions, Functors, and Combinators

Projection functions, *functors*, and *combinators* are the Rosetta constructs used to define interactions. Projection functions transform a facet into another form, such as a calculated value. They are used to project values from one domain to another. Functors are specialized projection functions that transform a facet into a new facet in a potentially different domain. They are used to move models between domains to facilitate composition, analysis, and reuse.

Finally, combinators take multiple facets and transform them into new facets. They are used specifically for facet composition and frequently involve product and sum operations.

Projection functions, functors, and combinators all treat Rosetta specification elements as data. They must examine the structure of facets and in most cases generate new facets. To facilitate this, Rosetta provides the special-purpose constructors, observers, and templates to observe and construct Rosetta abstract syntax elements (described in Chapter 12). Projection functions, functors, and combinators all examine the structure of a facet using observers and create new facet abstract syntax elements using constructors and template expressions. Thus, the same syntactic and semantic structures used for domain specification are used for facet manipulation.

15.1.1 Defining Projection Functions

A *projection function* moves information from one domain to another. All projection functions transform a source facet and a destination facet into a collection of terms in the domain of the destination facet. Effectively, the projection function moves information from the source domain into the destination domain specific to the two facets involved. Specifically, the signature of a projection function has the following form:

$$P(S::D_S,D::D_D)::\text{term\_list};$$

where P names the projection function. Here S is the source facet from domain D_S and D is the destination facet from domain D_D; both must be facets, but the associated parameter type may be a subtype of the general facet type. A projection function is in all ways a traditional Rosetta function and is defined using the same techniques. What distinguishes a project function is the presence of a facet type in the argument list.

EXAMPLE **15.1**

A Projection Function

The transform_terms projection function takes a facet from the static domain and transforms its terms into terms appropriate for use in the state_based domain by asserting that each is true in any state. The helper function transform_term takes a single term and asserts that it is true for all states using a universal quantifier and the ticked expression at:

```
transform_term(x::term)::term is
  << term forall(st::state |'x' 'at(st)) >>
```

The transform_terms projection function uses the transform_term function to transform its collection of terms into terms for the state_based domain:

```
transform_terms(F::static)::set(term) is
  image(transform_term,terms(F));
```

15.1.2 Defining Morphisms and Functors

While a projection function usually moves information from one facet to another, a *morphism* is a function that generates an entirely new facet. In addition to generating new terms, a morphism is responsible for generating parameter lists, variable sets, and export lists, and for selecting a domain for the new facet. Where a projection function moves information from one facet to another, a morphism generates one facet from another.

A *functor* is a special morphism that can operate on any facet in a domain. While morphisms may be specific to a particular facet, a functor maps every facet from one type into a facet from another. In essence, a functor generalizes a collection of morphisms to operate over a collection of facets. Functors are heavily used when moving specifications up and down the domain semi-lattice as well as when moving them between arbitrary domains.

One definition mechanism for morphisms is to define functions that construct new facets using template expressions. Using this approach, a traditional function is written that fills in a template expression, much like a constructed type to generate the new facet. The template expression would have the following form:

```
<< complete_facet_declaration facet 'name' 'signature' is
     'exports'
     'declarations'
   begin
     'terms'
   end facet 'name'
>>
```

where *name*, *signature*, *exports*, *declarations*, and *terms* are replaced by functions that generate the new abstract syntax or will be processed as a part of the template expression.

The second mechanism for constructing a new facet calls the **facet** constructor directly. The previous facet can be defined directly as:

```
make_complete_facet_declaration(name,signature,exports,declarations,terms)
```

The result is exactly the same AST structure as is generated by the template. The template example uses the template expression to take advantage of concrete syntax, with the advantage being a definition that is frequently easier to read. The function example uses the facet constructor to avoid the template expression completely, preferring to construct each AST structure. The choice is purely stylistic. When most of the elements of a construct are generated with functions, using the constructor is most appropriate. Otherwise, the template approach is preferred.

EXAMPLE 15.2

Functor for Moving
static Facets into the
state_based Domain

The gamma functor defined below uses the translate_terms function to translate a static facet into a state_based facet. The template for creating facets is instantiated for this specific case. Declarations and exports remain the same in the new facet. Terms are generated from the original terms using the previously

defined translator function. Finally, a new name and signature must be defined
for the new facet.

```
gamma(F::static)::state_based is
  let new_name :: facet_name be mangle(name(F));
      new_sig :: facet_signature be convert_sig(signature(F)) in
    << complete_facet_declaration facet 'new_name' 'new_sig' is
        'exports(F)'
        'declarations(F)'
      begin
        'translate_terms(F)'
      end facet 'new_name'
    >>
  end let;
```

The utility functions `mangle` and `convert_sig` are defined to create a new facet
name and a new signature, replacing `static` with `state_based` as the facet type.
Their complete definitions are omitted for brevity.   ■

15.1.3 Defining Combinators

A *combinator* is a functor that takes two facets and transforms them into a third
facet. The purpose of any combinator is to compose specifications for analysis
or further refinement. Combinators are defined much like morphisms, with the
exception that they take two facet arguments. Like morphisms, combinators can
be defined using template expressions and direct calls to facet constructor functions. Using sums and products of facets provides a third mechanism in addition
to templates and constructors.

The primary use for combinator is to provide a capability for composing and
refining or abstracting two facets in a single step. For example, a combinator could
be used to take a power consumption specification and a functional specification
for a digital component and generate a single simulation model that examines
both. This can be done by using one or more functors and a facet product, but
these operations are so common in Rosetta specifications that using the combinator to encapsulate the operations into a functor can simplify and improve the
specification.

EXAMPLE 15.3

Combinator for
Composing
discrete_time
Specifications into
Simulatable Models

A simple example of a combinator is the concretization combinator that composes two `discrete_time` facets into a single discrete event simulation (des) facet.
The operation simply uses the concretization function, gamma, from the interaction between `discrete_time` and des to move each model to the des domain.
Then a product operation composes them into a single model:

```
gammaC(x::discrete_time y::discrete_time)::des ;is
  discrete_time_des.gamma(x) * discrete_time_des.gamma(y);
```

The effect of this combinator is to compose two `discrete_time` models into a
single simulatable model. This combinator is useful for composing requirements

models and functional models to determine if requirements are met by an implementation. ∎

15.2 Defining Interactions

Rosetta interactions are less monolithic "things" than they are collections of information that define moving information between domains. Interactions are defined using a syntactic form that packages the various mechanisms for domain interaction. As such, each interaction defines translators, functors, and combinators between two domains. The syntax for defining an interaction is as follows:

```
interaction name(parameters) between D₁ and D₂ ⟦ as D₃ ⟧ is
   ⟦ export exportList | all ⟧;
   localDecls
begin
   begin translators
      ⟦ translatorDecls ⟧
   end translators;
   begin functors
      ⟦ functorDecls ⟧
   end functors
   begin combinators
      ⟦ combinatorDecls ⟧
   end combinators;
end interaction name;
```

In this definition, D_1 and D_2 are the source domains for models. The optional D_3 value is the common supertype of D_1 and D_2 for forming products and sums using the $*$ and $+$ operation. The default value for D_3 is the greatest common supertype of D_1 and D_2. It is possible that $D_1 = D_2$ in situations where translators must lift values from included facets into facets of the same type. Functors may also be written in such interactions, but the identity functor will suffice in most cases. Parameters defined for an **interaction** are of kind **design** similar to package parameters. Interaction parameters are used like generics to configure a general-purpose interaction for a specific use. However, interaction parameters are rarely used in practice.

All items defined an interaction's **translators**, **functors**, and **combinators** blocks are implicitly exported, the only exception being locally defined items appearing in the declarative region of each block. Such declarations must be constants and cannot be exported. Items declared in the interaction's declarative region are exported. The similarity between **interaction** definitions and **package** definitions is not accidental: both encapsulate related definitions. Where declarations in packages may be related for any number of reasons, declarations in interactions always define information flow between their associated domains.

15.2.1 Translators

> $name(parameters)$ **from** $x::T_{src}$ **in** D_{src} **to** T_{dest} **in** D_{dest}
> ⟦ **is** *expression* ⟧
> ⟦ **where** *expression* ⟧;

D_{src} is the source domain and D_{dest} is the destination domain; both must be either D_1 or D_2 defined in the interaction header. T_{src} and T_{dest} are the source and destination types; they will frequently be the same. The x parameter is for use in the translator definition, and *expression* is an expression of type T_{dest} that transforms x into a new value defined in D_{dest}. The value of an translator is not a function value, but defines a transformation to perform on parameters as a tick operator.

The **is** and **where** classes behave as they would in traditional function definitions. Both can be excluded to define a variable translator or translator signature. Thus, the signature of a translator can be defined without providing additional details.

15.2.2 Functor Definitions

> $name(parameters)$ **from** $x::D_{src}$ **to** D_{dest}
> ⟦ **is** *expression* ⟧
> ⟦ **where** *expression* ⟧ ;

Functor definition is nearly identical to translator definition, except that the x parameter is a facet of type D_{src} rather than a value defined in that domain. The functor is in all ways a function, with syntactic sugar added to the definition to enhance readability and to help correct functor definition.

15.2.3 Combinator Definition

> $name(parameters)$ **from** $x::D_{src_1}$ **and** $y::D_{src_2}$ **to** D_{dest}
> ⟦ **is** *expression* ⟧
> ⟦ **where** *expression* ⟧ ;

Combinator definition is nearly identical to functor definition, except the x and y parameters represent facets from the two source domains rather than from a single source domain. The combinator is in all ways a function, with syntactic sugar added to the definition to enhance readability and to correct combinator specification.

It should be noted that functors and combinators can be defined outside the structure of an interaction definition. They are simply functions that operate on language elements rather than on traditional data. It is perfectly reasonable to have a function with the signature:

```
morph(f::state_based)::finite_state;
```

defining a function of type:

```
<* (f::state_based)::finite_state *>
```

All operations on functions apply equally to functors and combinators, including currying and composition. Because translators are not traditional functions, such operations cannot be legally applied to them. For the same reason, translators are not easily defined outside the scope of an interaction. They can be defined as a part of functors and combinators as well as **with** clauses in domain definitions.

15.3 Including and Using Interactions

It is unwieldy to specify interaction elements whenever they are used in a specification. This is particularly true of translators that can clutter interfaces and reduce readability. Thus, the **use** clause defines default interactions for a given specification. An interaction is specified in a **use** clause in the same manner as a package is. The interaction is named and parameters are specified when required:

use $name(p_0, p_1, ..., p_n)$;

The **use** clause identifies specific interaction definitions in the same manner as for packages, using the dot notation to identify the library where the interaction exists:

use $p_0.p_1...name(v_0, v_1, ..., v_n)$;

where p_k are package or library names and v_k are values for interaction parameters.

A specific element of an interaction can be used by identifying it in the **use** clause:

use $p_0.p_1...name(v_0, v_1, ..., v_n).n$;

where n is the name of a functor, translator, or combinator function defined in the **interaction**.

When an interaction is used by a package, the **export** clause defined in the interaction controls visibility in the same manner as for a package. The only distinction is that for translators, the translator domains are used to select a translator when none is explicitly defined. Specifically, if a facet $F_{src}::D_{src}$ is included in a facet $F_{dest}::D_{dest}$ and translators are not specified for parameters of F_{src}, the domains D_{src} and D_{dest} select the translator. If a single translator is visible that translates from D_{src} to D_{dest}, then that translator is used. If multiple translators are visible, then the user must disambiguate in the specifications.

Default interactions are specified by use clauses at the package level. If a use clause appears prior to a package definition and references an interaction, then the interaction becomes the default for all facets, packages, domains, and components defined in the package. The default is overridden by using a specific interaction in conjunction with definitions in the package. Defaults for all declarations are defined at the outermost working package level.

15.3.1 Translator Usage

Translators are applied to facet parameters when one facet instantiates another using the attribute notation. The translator decorates the actual parameter. The specific notation is:

   label : *name*($p't,...$);

where *name* is the instantiated facet name, *p* is an actual parameter instantiating the associated formal parameter, and *t* is the translator function used to move information from the included facet domain to the including facet domain. It is possible for *t* to have parameters, but all parameters must be specified when the translator is used.

Whether the formal parameter associated with *p* is an input or output parameter, instantiating it with $p't$ asserts that the application of *t* to *p* must result in a value compatible with constraints on the formal parameter.

15.3.2 Functor and Combinator Usage

Functors are simply functions whose domain and range include facets. Thus, functor application is function application. Specifically:

   name($p_0, p_1, ..., F$)

evaluates to the application of functor *name* to a facet F. The facet is a required parameter. Other specified parameters, p_0, ..., p_k, precede the facet parameter in the argument list. This is done to allow currying to specialize functors.

Combinators are also functions whose domain and range include facets. The only distinction from functors is that the combinator must include two facets. Like functor application, combinator application is identical to function application. Specifically:

   name($p_0, p_1, ..., F_1, F_2$)

elaborates to the application of the combinator *name* to facets F_1 and F_2. The facets are required parameters. Other specified parameters, $p_0, ..., p_k$, precede the facets parameters in the parameter list. Again, this is done to allow currying to specialize combinators.

15.4 Existing Rosetta Interactions

Two types of interactions exist in the base Rosetta system. *Semi-lattice* interactions are defined between interconnected domains in the domain semi-lattice. They define the abstraction and concretization functions alpha and gamma that allow moving a specification in the semi-lattice. In addition, semi-lattice interactions are created whenever a new domain is added.

Ad hoc interactions are defined by hand when an interaction between domains is desired. These interactions provide mechanisms for moving directly between domains when it is more natural to be moving up and down the semi-lattice. User-defined ad hoc interactions may also define interactions between domains when default interactions do not suffice.

15.4.1 Semi-Lattice Interactions

Whenever a domain is included in the domain semi-lattice, an interaction exists between it and its supertype and subtypes. The interaction between a domain and its supertype is given the name alpha and represents an abstraction of the domain into its supertype. Specifically, alpha(f) moves f from its current domain to its immediate supertype domain. The interaction between a domain and its subtypes is given the name gamma and represents a concretization of the domain into one of its subtypes. Specifically, gamma(f) moves f from its current domain to one of its subtypes, as determined by ascription or context.

Instances of both alpha and gamma are calculated from the extension that refines a domain into its subtype. In Chapter 12, the **where** clause is defined in domains to specify how a facet in a domain is elaborated. Specifically, given f::D, the **where** clauses define how D is extended with elements from f to define a facet. Because a domain extends its supertype like a facet extends its type, the **where** clauses also define the instance of gamma needed to concretize a specification, moving it from its type to one of its subtypes.

The inverse of the transformation defined by **with** clauses defines the instance of alpha used to abstract a specification, moving it from its type to its supertype. If a specification is concretized using gamma and abstracted using the related alpha, the result is the original facet in the abstract domain:

```
alpha(gamma(f)) == f
```

Unfortunately, the same relationship does not hold when abstracting a specification. If the original specification uses items defined in the concrete domain, those definitions will be lost in the abstraction and cannot be regenerated. When applying gamma, we know exactly what information is added to a specification. When defining a specification in the concrete domain, we cannot make the same claim. A specific case of this occurs when applying alpha and gamma to move between discrete_time and infinite_state. If specific time values are used in a discrete_time specification, such as:

```
x@(t+1) = f(x)
```

there is no way to preserve the time value in the infinite_state specification. The information is necessarily lost. However, if the discrete_time specification uses next rather than a specific time reference, applying alpha results in a specification than can be concretized:

```
x@next(t) = f(x)
```

The next function exists in infinite_state, but simply at a more abstract level.

The implication is that while gamma can always be applied to make a specification more concrete, alpha is restricted. Specifically, if a specification uses items defined in the concrete domain, alpha may not produce the desired abstraction. If specification reuse composition is a goal, such relationships must be taken into consideration.

15.4.2 Ad Hoc Interactions

Ad hoc interactions exist between many domains that are not related by a subtype relationship. For example, moving between the frequency and continuous_ time domains or the finite_state and discrete_time domains requires information beyond that defined by semi-lattice extensions. Unlike those generated automatically by adding domains to the semi-lattice, ad hoc interactions are always written by hand.

EXAMPLE 15.4

Using alpha and gamma
for Moving within the
Semi-lattice

Using alpha and gamma interactions to move between domains has great utility when composing specifications. For example, in activity-based power modeling, power consumption is defined over state change. When a system's state changes, power consumption is estimated based on the type of state change. Thus, it makes good sense to define power models in the state_based domain:

```
facet power
  [Ti,To::type]
  (x::input Ti; y::output To;
   leakage::design real;
   change::design real)::state_based is
  export p;
  p::real;
begin
  p' = p + leakage + if event(y) then change else 0 end if;
end facet power;
```

As a system specification involving the power facet is refined, the specifics of state may change substantially. Ideally, the power model should be refined along with the system model by default. Thus, if we write a model in the discrete_time domain, then we could refine the power model to exist in the discrete_time domain:

```
power_dt[Ti,To::type](x::input Ti; z::output To;
        l,c::design real) :: state_based is
  gamma(gamma(power(x,z,l,c)));
```

Similarly, alpha can move a specification up in the domain semi-lattice. Looking at power aware modeling again, it may be more desirable to abstract the system model to the state_based domain for evaluation. In the same manner that power moved down, system can move up:

```
system_sb(x::input bit; z::output bit) :: state_based is
  alpha(alpha(system(x,z)));                                        ∎
```

An excellent example of an ad hoc interaction occurs between the continuous_time and frequency domains. Mathematically, this interaction defines the Fourier transform and inverse Fourier transform moving between frequency and time. This interaction is predefined; named fourier_transform, it defines fourier and inverse_fourier morphisms. The signature of this interaction has the following form:

```
interaction fourier() between continuous_time and frequency is
  export all;
begin
  begin translators
  end translators;
  begin functors
    fourier() from continuous_time to frequency;
    inverse_fourier() from frequency to continuous_time;
  end functors;
  begin combinators
  end combinators;
end interaction fourier;
```

Only functors are defined in this interaction, allowing a user to specify the Fourier and inverse Fourier equivalents of a facet. However, it is not possible to interconnect or compose continuous_time and frequency domains using pre-existing functions. Specifiers are of course allowed to write their own translators and combinators if desired.

Although an interaction between state_based and static domains is generated automatically by the semi-lattice construction, defining the interaction by hand provides useful insight. The *ad hoc* definition in Figure 15.1 defines the translators, functors, and combinators necessary to combine state_based and static facets.

Two translators facilitate moving information through parameters between domains. The alphaT translator moves information from state_based to static while gammaT moves information from static to state_based. Moving from static to state_based adds the 'at(s) notation, to add dereferencing to the static parameter. Moving from state_based to static simply replicates the actual parameter.

Two combinators compose specifications by moving one to a higher or lower abstraction level. The alphaC combinator moves its state_based argument into the static domain and performs composition. Conversely, the gammaC combinator moves its static argument into the state_based domain and performs composition.

Finally, the two functors alpha and gamma move specifications between domains. Specifically, alpha moves a specification from the more concrete domain,

```
interaction static_state_based() between static and state_based is
  export all;
begin
  begin translators
    alphaT[T::type]() from x::T in state_based to T in static is
      if is_tick_operation(x) then operand(x) else x end if;
    gammaT[T::type]() from x::T in static to T in state_based is
      << tick_operation 'x' 'at(s) >>;
  end translators;
  begin functors
    alpha() from state_based to static;
    gamma() from  static to state_based;
  end functors
  begin combinators
    alphaC() from x::state_based and y::static to static is
      alpha(x) + y;
    gammaC() from x::static and y::state_based to state_based is
      gamma(x) + y;
  end combinators;
end interaction static_state_based;
```

Figure 15.1 An interaction defining the relationship between the state_based and domains.

state_based, to the more abstract domain, static. The functor gamma performs the inverse, moving specifications from static to state_based. The details of these functors are omitted, as they will be generated automatically by the construction of the semi-lattice.

15.4.3 Composite Interactions

A common mechanism for defining new interactions is to compose elements from existing interactions. Because functors and combinators are simply functions, it is easy to define new interactions by composing interaction elements using function composition. Figure 15.2 defines such an interaction between finite_state and discrete_time domains. Such an interaction would be useful when taking a finite state machine and moving to discrete time for simulation or integration with a larger system.

Moving between finite_state and discrete_time is a simple matter of applying multiple alpha and gamma functors. Defining the interaction simply institutionalizes the interaction and increases readability. Each functor and combinator is defined by composing existing functions using the function composition operator.

The finite_state_discrete_time interaction defines two functors and three combinators (Figure 15.3). The functors are defined by composing alpha and gamma functions from interactions defining paths between domains in the semi-lattice. Both functors move a specification from one domain to the other defined in the interaction.

The first two combinators are similar to the functors in that they define moving one specification to the other's domain and performing a facet product. The functors defined in the interaction are used to define the combinator. The third combinator

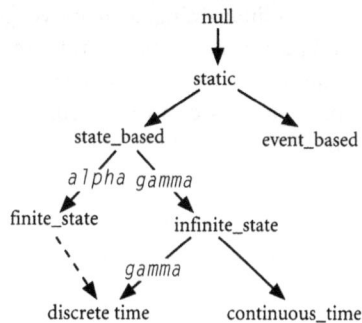

Figure 15.2 The domain semi-lattice before and after creation of a finite_state to discrete_time interaction.

```
use state_based_finite_state(),
    state_based_infinite_state(),
    infinite_state_discrete_time();
interaction finite_state_discrete_time() between
finite_state and discrete_time is
begin
  begin translators
  end translators;
  begin functors
    finite_state_to_discrete_time() from finite_state to discrete_time is
      infinite_state_discrete_time.gamma.
      state_based_infinite_state.gamma.
      state_based_finite_state.alpha;
    discrete_time_to_finite_state() from discrete_time to finite_state is
      state_based_finite_state.gamma.
      state_based_infinite_state.alpha.
      infinite_state_discrete_time.alpha;
  end functors;
  begin combinators
    finite_stateC() from x::finite_state and y::discrete_time to
      finite_state is
      x * discrete_time_to_finite_state(y);
    discrete_timeC() from x::finite_state and y::discrete_time to
      discrete_time is
      finite_state_to_discrete_time(x) * y;
    state_basedC() from x::finite_state and y::discrete_time to
      state_based is
      state_based_finite_state.alpha(x) *
      (state_based_infinite_state.alpha . infinite_state_discrete_time)(y)
  end combinators;
end interaction finite_state_discrete_time;
```

Figure 15.3 Interaction between finite_state and discrete_time, constructed by composing interaction components.

is more interesting, as it moves both specifications into the state_based domain and performs composition there. In effect, both specifications are moved to a common abstract domain and are composed. This combinator might be used to perform analysis or to lift results from a concrete specification into a more abstract context.

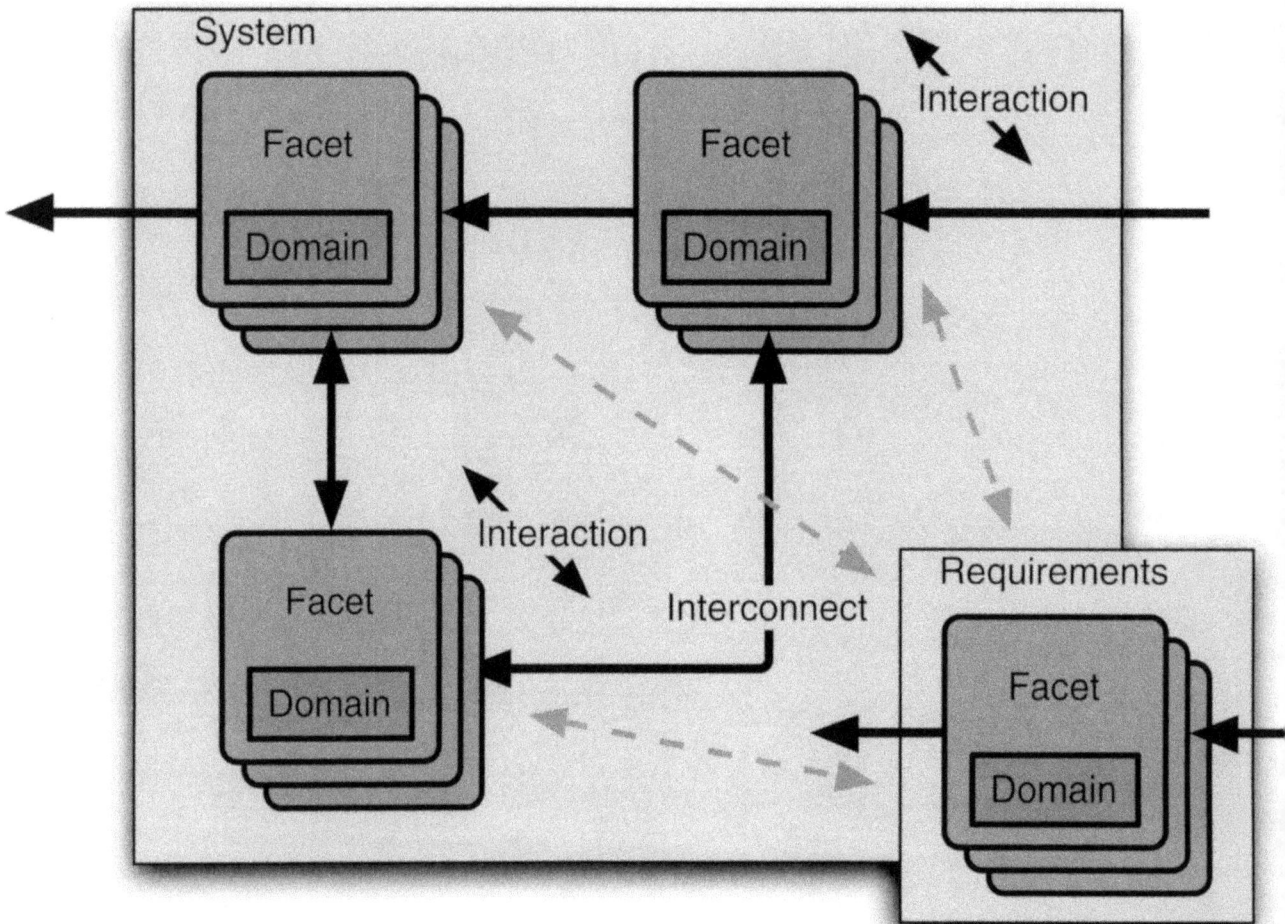

V

Case Studies

The *case studies* provide a collection of usage examples for Rosetta specifications. They are intended to demonstrate common specification techniques and structuring styles for different specification types.

After completing the chapters in Part V, you will have seen four specifications of system level properties in RTL design, power-award design and network access control.

16 Case Studies

The best way to understand system-level modeling is to model systems. Thus, several case studies are presented in Chapters 17–20 to help illustrate some of the concepts seen thus far and integrate smaller examples into larger systems. The first case study looks at traditional RTL design in Rosetta. It is more a primer than a true case study, showing how to structure designs and providing a jumping-off point from a traditional domain. The RTL case study looks at the design of a traditional digital system from both a structural and a behavioral level. It shows a methodology for packaging designs and for specifying correctness properties.

The second and third case studies examine power-aware modeling in digital systems design. In this classic example, models are developed for function, power consumption, and power constraints. They are then composed to define a constrained system-level model. Finally, the power consumption model is refined three ways to represent power consumption in CMOS, field programmable gate array (FPGA), and software implementations.

The final case study examines system-level requirements of access control across communications networks. It models how a portable system accesses protected services as it moves through different network infrastructures. One infrastructure represents a protected, local network where controlled resources reside. Another infrastructure represents a public network with no protections or assurances. Access control requirements are invariant, yet the mobile system would like to access requirements regardless of the network infrastructure being used.

16.1 Methodology

Although the case studies examine different domains, the methodologies used in their development share a common pattern. First, modeling goals are established and modeling domains selected. This provides a structure or anatomy for the specification. Next, models are written for basic system elements in selected domains. Functors, products, and co-products are then used to construct composite specifications. Finally, combinators are used to generate analysis and synthesis models.

16.1.1 Identify System-Level Modeling Goals

One of the greatest myths of system-level design is the belief that analysis goals can be identified after models are constructed. This is akin to asserting that you'll know what you need when you see it. This *post hoc* approach has never and will never result in successful design activities where complex systems are involved. System-Level analyses goals must be reflected in the specification.

System-level modeling goals must reflect the overall system design goals. Our case studies involve three different domains that require three different modeling architectures. Our first system's primary requirement is correctness. Thus, models will concentrate on representing functional requirements similar to existing hardware description language (HDL)-based design processes. Our second system's primary requirement is again functional correctness, but in the context of a power constraint. Thus, the models not only reflect correctness conditions, but account for interactions between functional requirements and performance constraints. The final system also involves performance constraints in the form of access control requirements. Here the objective is not defining a system, but exploring implications of a system design.

In each case study, the models developed reflect design goals as necessitated by the complexity of system-level design. Rosetta provides modeling support for each activity as reflected by the case studies. In each case, heterogeneity and model composition play a dominant role in the modeling activity.

16.1.2 Identify Basic Models and Domains

Knowing what the analysis goals are leads to identifying the necessary system models and their associated domains. The key is identifying the best domain for each model. Two equally important observations play key roles here: (i) the most natural domain for each model and (ii) relationships between domains in the domain semi-lattice. Using the most natural domain is the ideal approach. However, if interactions do not exist between domains, or if existing interactions are weak, analysis can be made less difficult by choosing strongly interrelated domains.

Full system-level models will virtually always contain models representing functions, effects on system resources, and constraints on system resources. Functional models typically represent requirements, or "what" the system is to do, and implementations, or "how" the system achieves its goals. Models of effects on system resources map functional behavior to impacts on constraints. For example, a power model would map functional behavior to power consumption, a security model would map functional behavior to resource access and usage, and a synthesis model would map functional behavior to utilization of computing resources. Finally, constraint models represent the resources available to the system. They represent the conditions the system must operate under. For example, a power constraint indicates how much power a system may consume, a security constraint indicates access control requirements on system resources, and a synthesis constraint indicates available computing resources.

In some cases, the most appropriate modeling domains will not exist in the domain semi-lattice. When this occurs, new domains and interactions must be constructed. This should always be done by extending an existing domain, even if the only domain applicable is the `static` domain. The easiest mechanism is defining an engineering domain. Adding new specification vocabulary to an existing engineering domain or to a model-of-computation domain is a reasonably straightforward process, typically involving writing new functions and defining types. If an engineering domain cannot be defined directly, then a model-of-computation domain may be defined by either extending an existing model-of-computation domain or refining a unit-of-semantics domain. Defining a new unit-of-semantics domain should be a last resort and only considered when dramatically new domains are required.

16.1.3 Define Basic Models

Now the modeling task begins. The systems designer must construct each of the basic models. Most of these models can be constructed in some degree of isolation. However, it is critical that points of interaction between models be considered. These include interfaces and visible quantities exported from each model. For example, when modeling power consumption, it is important to think about how power will be referenced from each appropriate model. Although the functional model will have no observable power property, the power constraint and power consumption models must. Furthermore, the observations must be represented to facilitate information exchange between models.

16.1.4 Construct Composite Models

With the basic models in place, interactions are used to compose them and perform the desired analysis. By choosing domains early in the design process with well-defined interactions between them, functors, translators, and combinators are available to compose models. If this is not possible, interactions must be defined for the specific design problem.

Choose the domain most appropriate for performing analysis. This may be the domain common to the majority of models or a domain for which excellent analysis tools are available. In most cases, this choice is made when the modeling domains are chosen and impacts the structure of the entire modeling process.

Use functors to move models to the analysis domains. If interactions between domains exist, then they are easily used individually and in composition to move

move models. If interactions between domains do not exist, then following this path requires the user to write their own interactions.

Use products to combine models whenever two or more models must be concurrently satisfied. Models should be composed when and where most appropriate. The composition does not need to wait until all models are in their final domains. Remember, this does not generate analysis models nor does it fully implement interactions between domains. It bundles models together, indicating they must be satisfied concurrently or in other combinations, as indicated by composition operators.

Use combinators to generate analysis models from product and co-product models. These combinators move composite models from modeling domains to domains associated with analysis tools. At this point, analysis is performed to begin understanding system-level effects of local design decisions. If the processes of model transformation are fully automated, then different design options and analysis criteria can be selected.

16.2 Before Proceeding

Before approaching the case studies, it is important to realize that each example has been developed over time. The power-aware modeling case study has been a standard in the Rosetta community since the earliest days of language requirements design. System-level designs do not happen the first or second time through. They involve significant planning, re-planning, modeling, and re-modeling. The original power-aware case study bears virtually no resemblance to the version presented here.

The natural question must now be, "if it's so resource interactive and difficult, then why do this?" The answer is quite simple. One can either discover problems with system-level decisions during system-level design, or later in the design process, when they are immensely more expensive to correct. What Rosetta enables is this process of making trade-off decisions and exploring design alternatives with predictive analysis early in the design process.

Register-Transfer-Level Design

The register-transfer-level design case study examines defining requirements and an architecture for a simple alarm clock controller. This is an appropriate case study because it uses common abstractions from electronic systems design and extends those abstractions to include elements of system-level design. The first model is exclusively behavioral and uses a single behavioral specification to define systems requirements. The second model uses a collection of behavioral specifications to define an architecture for the controller. Finally, some correctness conditions are defined using Rosetta's special-purpose component constructs.

Figure 17.1 graphically defines both the system-level view and the architecture to be defined. Defined in terms of signals from Figure 17.1, the requirements for the clock controller are:

When the setTime bit is set, (timeInHr*60 + timeInMin) is stored as the clockTime while timeInHr and timeInMin are output to the time display.

When the setAlarm bit is set, (timeInHr60 + timeInMin) is stored as the new value of alarmTime while timeInHr and timeInMin are output to the time display.

When the alarmToggle bit is set, the alarmOn bit is toggled.

When clockTime and alarmTime are equivalent and alarmOn is asserted, the alarm should be sounded. Otherwise it should not.

When setTime is clear and setAlarm is clear, clockTime is output as the display time.

The clock increments its time value when clk is asserted.

The models defining this case study are organized into three collections. The timeTypes package defines basic types and functions used throughout both the behavioral and structural models. The alarmClockBehav model defines functional requirements for the clock system. Each of the previous requirements maps into one or more requirements in the alarmClockBehav model. Several RT level specifications define the components of the implementation architecture defined by alarmClockStruct. These include a data store, multiplexer, comparator, and

285

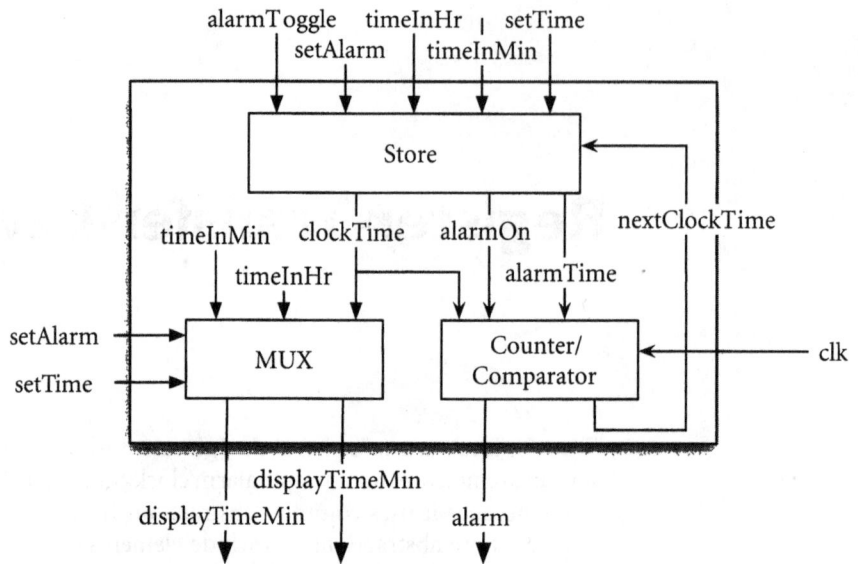

Figure 17.1 System-level architecture of a simple alarm clock controller.

counter defined as individual models. Finally, alarmClock defines relationships between the behavioral and structural alarm clock models that define correctness of the architecture.

Each model in this case study prototypical of models useful when designing using RTL components. The requirements model provides a behavioral description of the entire system of interest. The component models provide excellent examples of RTL specifications in Rosetta, while the structural model exemplifies structural assembly. Finally, the requirements model shows how correctness conditions can be expressed as a part of the system model.

17.1 Requirements-Level Design

Two specifications define the requirements-level models for the alarm clock controller. The first is a package of types and functions used across all models. The timeTypes package represents common definitions that are shared or appear in the interfaces of specifications. The second is a model defining the system-level requirements. The alarmClockBehav model translates the English description from the introduction into a formal Rosetta specification.

Figure 17.2 shows the timeTypes package specification. The only element of this package is a function for incrementing time values that is used frequently

```
package timeTypes::static is
   increment_time ( minutes :: integer ) :: integer is
      if minutes < 12 * 60
          then minutes + 1
          else 0
      end if;
end package timeTypes;
```

Figure 17.2 Time types shared among specifications.

```
use timeTypes;
facet interface alarmClockBehav
        ( timeInHr, timeInMin :: input integer;
          setAlarm, setTime, alarmToggle, clk :: input bit;
          displayTimeHr, displayTimeMin :: output integer;
          alarm :: output bit ) :: state_based is
end facet interface alarmClockBehav;
```

Figure 17.3 The interface of the system-level requirements model.

throughout the case study. Frequently, a package like this will contain definitions of types as well as functions. This is particularly useful for shared types used in interfaces that cannot be defined within facets. Because interface types are standard Rosetta base types, this is not necessary here.

Figures 17.3 and 17.4 show the basic requirements specification, similar to what would be called a behavioral specification in VHDL or Verilog. This specification is separated into interface and body to simplify the presentation. The interface specification defines the signals that appear at the system interface described in Figure 17.1. The association between model parameters and signals from the figure is readily visible.

The specification body takes the requirements specified in the introduction and formalizes them as specification terms. Each term label describes the action specified by that label. In each case, **if** expressions or implications define the relationship between inputs and current state, and the values to be stored or output in the next state. Remember that terms are simultaneously true and their order within the model is immaterial.

A deeper explanation of several terms will help to understand the specification and its style. setClock_label is the name of the term specifying how and when the clock time gets set. There are three possible cases defined in this term. When setTime is asserted, the internal clock time is set to the time inputs and the input time is displayed. This corresponds to behavior when setting a digital clock. If the clock is not being set and there is a rising edge on clk, then the internal state is updated by incrementing the current state. Finally, if nothing has happened, then the internal clock state does not change. The setAlarmLabel_term is similar, without the need to update the internal clock.

```
use timeTypes;
facet body alarmClockBehav is
 alarmTime, clockTime :: integer alarmOn :: bit;
begin
  setClock_label:
    if %setTime
      then(clockTime' = timeInHr * 60 + timeInMin)
           and(displayTimeHr' = timeInHr)
           and(displayTimeMin' = timeInMin)
      elseif event(clk) and clk=1
           then clockTime' = increment_time(clockTime)
      else clockTime' = clockTime
    end if;
  setAlarm_label:
    if %setAlarm
      then(alarmTime' = timeInHr * 60 + timeInMin)
           and(displayTimeHr' = timeInHr)
           and(displayTimeMin' = timeInMin)
      else alarmTime' = alarmTime
    end if;
  displayClock_label:
    ((setTime = 0 )and(setAlarm = 0)) =>
      (displayTimeHr' = clockTime div 60)
      and(displayTimeMin' = clockTime mod;60);
  armAlarm_label:
    if event(alarmToggle) and alarmToggle = 1
      then (alarmOn' = not alarmOn)
      else (alarmOn' = alarmOn)
    end if;
  sound_label:
    alarm' == if (alarmOn = 1)and(alarmTime = clockTime)
                 then 1;
                 else 0
              end if;
end facet Body alarmClockBehav;
```

Figure 17.4 System requirements specification.

The clock's display is managed by the displayClock_label term. If the clock is not being set and the alarm is not being set, this term specifies that the clock display should be the current time. Note that this term is not dependent on the clock and does not reference any variables in the next state. Thus, the term must always hold for the clock to function properly. If nothing is being set, the clock must always display the current time.

The armAlarm_label and sound_label manage the arming and sounding of an alarm. armAlarm_label states that the alarm state is toggled when a rising edge occurs on the alarmToggle signal. Note that this is an asynchronous signal, as the term does not reference the clock. sound_label indicates that the alarm is sounded whenever the stored alarm time is equal to the stored clock time. Like the display requirement, this requirement does not depend on the clock or reference variables in the next state. Thus, the term is invariant across all states.

As usual, symbols decorated with ticks represent values in the next state. In most cases, the use of ticked symbols is limited to the left side of equalities. This is in no way required by Rosetta. However, it reflects a common HDL style, where the next value to be assigned to a signal is calculated and a signal assignment operator used to schedule an update. Specifications written in this form are easier to synthesize and simulate. However, when writing specifications of incomplete systems, this style may not be appropriate or even feasible.

17.2 Basic Components

The basic component set defines a collection of custom RTL components that will be used to define a structural architecture for the controller. One facet is defined to implement each of the basic functional blocks from Figure 17.1. These facets will be assembled later to define the complete architecture.

17.2.1 Multiplexer

The mux component in Figure 17.5 is a specialized multiplexer that selects from three different times to display. Two control signals, setAlarm and setTime, are used to choose whether to display the new alarm time or clock time being entered, or the time maintained by the clock. Where as the store component will use these signals to control when data is replaced in the internal clock store, the mux component uses them to determine what value should be displayed.

```
use timeTypes;
component mux ( timeInHr, timeInMin :: input integer;
                clockTime :: input integer;
                setAlarm, setTime :: input bit;
                displayTimeHr, displayTimeMin :: output integer
              ) :: state_based is
begin
  assumptions
    ambiguous_select: not(%setTime and %setAlarm)
  end assumptions;
  definitions
    l1: %setAlarm or %setTime =>
          (displayTimeHr' = timeInHr)and(displayTimeMin' = timeInMin);
    l3: not (%setTime or %setAlarm) =>
          (displayTimeHr' = clockTime div 60)
          and (displayTimeMin' = clockTime mod 60);
  end definitions;
  implications
  end implications;
end component mux;
```

Figure 17.5 The clock multiplexer component.

A component is used for this model, rather than a facet, because of the desire to specify a usage condition. The ambiguous_select term in the **assumptions** section asserts that this component requires that these two inputs must not be simultaneously asserted. If so, the function of the device cannot be uniquely determined. Using the assumption in this way simplifies the component specification by eliminating the need to specify behavior for this case. It is simply deemed a bad input state that cannot be managed.

17.2.2 Data Store

The store component in Figure 17.6 defines a simple register specialized for storing time values used in this system. The device is rising-edge triggered on any one of three signals. The state_based predicate event checks to see if a value has changed during the state change. When conjuncted with a check to see if the new value is 1, the result is a rising-edge check. In effect, store behaves like three registers packaged in the same component.

Possibly the most interesting thing about store is that it uses no internal variables to store values. Because each value is associated with its own update signal and the stored value is the same as the output value, the output parameter is effectively the store. The model could have as easily represented the internal store explicitly.

A facet model is used for the store component because there are no correctness conditions or usage assumptions associated with the device. A component could easily be used, but the assumptions and implications blocks would be empty. The only reason to use a component would be in anticipation of additional information or back annotation during analysis.

```
use timeTypes;
facet store ( timeInHr, timeInMin, nextClockTime :: input integer;
              setAlarm, setTime, alarmToggle :: input bit;
              clockTime, alarmTime :: output integer;
              alarmOn :: output bit ) :: state_based is
begin
  l1: if event(setAlarm) and %setAlarm
        then alarmTime' = timeInHr * 60 + timeInMin
        else alarmTime' = alarmTime
      end if;
  l2: if event(setTime) and %setTime
        then clockTime' = timeInHr * 60 + timeInMin
        else clockTime' = clockTime
      end if;
  l3: if event(alarmToggle) and %alarmToggle
        then alarmOn' = not alarmOn
        else alarmOn' = alarmOn
      end if;
end facet store;
```

Figure 17.6 The alarm clock store component.

17.2.3 **Counter and Comparator**

The counter specification (Figure 17.7) is somewhat misnamed, as it defines the next state function for the internal timer, but does not store it. Instead, the value is output to the store component, where it is saved. In effect, the counter outputs a new value each time the clk signal rises and outputs the current input otherwise.

The comparator specification (Figure 17.8) implements an equality comparison on time values and outputs the result if the clock's alarm is engaged. In effect, it is a comparator whose output is conjuncted with the setAlarm input.

17.2.4 **Clock**

The clock facet in Figure 17.9 defines a signal that simply changes state. The clock is defined in the state_based domain, implying that nothing can be said about its period. It is modeled simply as a component that inverts its output in each state

```
use timeTypes;
facet counter ( clockTime :: input integer;
                clk :: input bit;
                nextClockTime :: output integer ) :: state_based is
begin
  14: nextClockTime' = if event(clk) and %clk
                          then increment_time(clockTime)
                          else clockTime
                       end if;
end facet counter;
```

Figure 17.7 Alarm clock counter component.

```
facet comparator ( setAlarm :: input bit;
                   alarmTime, clockTime:: input integer;
                   alarm :: output bit ) :: state_based is
begin
  11: alarm = alarmOn and %(alarmTime = clockTime);
end facet comparator;
```

Figure 17.8 Alarm clock comparator.

```
facet clock(x::output bit)::state_based is
begin
  update: x' = if x=0 then 1 else 0 end if;
end facet clock;
```

Figure 17.9 A state_based clock component.

change. Note that in this case the model is a facet rather than a component, due to the simplicity of the device. There is no need to understand usage assumptions or specify correctness conditions for the clock.

17.3 Structural Design

With the component models defined, the structural design merely provides interconnections between components. Like the behavioral specification, the structural specification is split into interface and body. The interface, shown in Figure 17.10, is uninteresting, with the exception of being identical to the behavioral interface (excluding the name of the model). Rosetta does not currently support multiple implementations for the same interface, thus the structural and behavioral interfaces must both exist.

The specification body, shown in Figure 17.11, provides interconnection details. Each internal component has a unique design and is thus instantiated only once within the structural design. A local declaration defines internal signals using the same naming conventions used in Figure 17.1.

```
use TimeTypes;
facet interface alarmClockStruct
  ( timeInHr, timeInMin :: input integer;
    setAlarm, setTime, alarmToggle, clk :: input bit;
    displayTimeHr, displayTimeMin :: output integer;
    alarm :: output bit
  ) :: state_based is
end facet interface alarmClockStruct;
```

Figure 17.10 Alarm clock structural implementation interface.

```
facet body alarmClockStruct is
    clockTime, nextClockTime, alarmTime :: integer;  alarmOn :: bit;
begin
  store_1 : store ( timeInHr, timeInMin, nextClockTime, setAlarm,
                    setTime, alarmToggle, clockTime, alarmTime, alarmOn );
  counter_1 : counter ( clockTime, clk, nextClockTime );
  comparator_1 : comparator ( setAlarm, alarmTime, clockTime, alarm );
  mux_1 : mux ( timeInHr, timeInMin, clockTime,
                setAlarm, setTime, displayTimeHr, displayTimeMin);
end facet alarmClockStruct;
```

Figure 17.11 Alarm clock structural implementation body.

17.4 Design Specification

At this point in the alarm clock design, requirements exist in the form of a behavioral specification, and architectural requirements exist in the form of a structural specification. However, there is no correctness condition explicitly stating the relationship that should exist for them. An implied correctness condition can be read into the system description. However, this is not sufficient for achieving true system-level design.

The component alarmClock shown in Figure 17.12 defines correctness between the structural and behavioral models in a component that can be used in other models. The alarmClock component instantiates a single copy of alarmClockBehav in its **definitions** section. By instantiating the behavioral specification with parameters from the component interface, the component now has the behavior specified by the behavioral specification.

The component's correctness condition is expressed in the **implications** section using the term:

```
correctness: alarmClockStruct(timeInHr,timeInMin,setAlarm,setTime,
                              alarmToggle,displayTimeHr,displayTimeMin,
                              alarm) =>
             alarmClockBehav(timeInHr,timeInMin,setAlarm,setTime,
                              alarmToggle,displayTimeHr,displayTimeMin,
                              alarm);
```

```
component alarmClock
      ( timeInHr, timeInMin :: input integer;
        setAlarm, setTime, alarmToggle, clk :: input bit;
        displayTimeHr, displayTimeMin :: output integer;
        alarm :: output bit
      ) :: state_based is
begin
  Assumptions
  end Assumptions
  definitions
     def: alarmClockBehav(timeInHr,timeInMin,setAlarm,setTime,
                          alarmToggle,clk,displayTimeHr,
                          displayTimeMin,alarm);
  end definitions;
  implications
     correctness: alarmClockStruct(timeInHr,timeInMin,setAlarm,setTime,
                          alarmToggle,clk,displayTimeHr,
                          displayTimeMin,alarm) =>
                  alarmClockBehav(timeInHr,timeInMin,setAlarm,setTime,
                          alarmToggle,clk,displayTimeHr,
                          displayTimeMin,alarm);
  end implications;
end component alarmClock;
```

Figure 17.12 Specification of design requirements and correctness conditions.

The => operation on facets and components is defined as homomorphism. Thus, the correctness condition asserts that all properties of the `alarmClockStruct` component must be derivable from the `alarmClockBehav` component. This is the same instantiation of the component specified in the **definitions** block.

The correctness condition accomplishes two important things. First, it explicitly states that all behaviors of the behavioral specification must be exhibited by the structural specification. This is a rather strong correctness condition, but for such a small component this is quite manageable. The second, and less obvious, result is that the behavioral specification is tied to the structural specification. If the structural specification is altered in a way that causes it to not satisfy the behavioral specification, the correctness condition is violated and one of the specifications must be modified. Thus, the behavioral specification tracks the structural specification with the component defining correctness wherever it is used.

An alternate definition of correctness is shown in Figure 17.13 as a component defining testing conditions. The `alarmClockTest` component instantiates one copy of the behavioral and structural specifications. It then drives inputs of each internal component with external signals from the interface.

Each component's outputs are gathered by a collection of internal variables. Instead of generating outputs, the `alarmClockTest` component compares like outputs from the behavioral and structural models using these variables. These assertions are made in the **implications** section and are invariant over state change. This is a weaker correctness condition than that defined earlier because it specifies only three properties that must be maintained. The name `alarmClockTest` is chosen because the component defines a testing architecture in addition to correctness conditions.

17.5 Wrap Up

The alarm clock controller is included as an example of how Rosetta extends classical RTL design. At the functional design level, Rosetta adds little over VHDL or Verilog. Components are defined using a largely operational subset of the expression language, similar in nature to behavioral VHDL or Verilog specification. Components are then integrated using structural constructs, again similar to structural VHDL or Verilog.

Where Rosetta introduces new capabilities is in the inclusion of usage assumptions and correctness conditions. In the components `alarmClock` and `alarmClockTest`, the **implications** and **definitions** sections are used to define usage assumptions and correctness conditions. Thus, design intent is memorialized as a part of the system specification in an interpretable fashion.

Recording usage assumptions formally as requirements and implications has two advantages. First, design intent is not lost and moves through design stages with component requirements. Design intent becomes a part of the living specification. Second, design intent is usable by tools in the design flow. Representing

```
component alarmClockTest
    ( timeInHr, timeInMin :: input integer;
      setAlarm, setTime, alarmToggle, clk :: input bit;
    ) :: state_based is
  displayTimeHrS, displayTimeMinS :: integer;
  displayTimeHrB, displayTimeMinB :: integer;
  alarmS :: output bit
  alarmB :: output bit
begin
  Assumptions
  end Assumptions
  definitions
    beh: alarmClockBehav(timeInHr,timeInMin,setAlarm,setTime,
                         alarmToggle,clk,displayTimeHrB,
                         displayTimeMinB,alarmB);
    struct: alarmClockStruct(timeInHr,timeInMin,setAlarm,setTime,
                         alarmToggle,clk,displayTimeHrS,
                         displayTimeMinS,alarmS);
  end definitions;
  implications
    c1: displayTimeHrS == displayTimeHrB;
    c2: displayTimeMinS == displayTimeMinB;;
    c3: alarmS == alarmB;
  end implications;
end component alarmClockTest;
```

Figure 17.13 Specification of correctness conditions in a testing component.

intent formally in Rosetta enables input and analysis using tools in the design flow. For example, both requirements and implications can frequently be used as assertions in simulators and model checkers. The designer is thus rewarded for maintaining this information as part of the specification.

Power-Aware Design

The first power-aware design study examines Rosetta's capabilities for transforming and composing models to examine power trade-offs with respect to implementation technologies. The challenge is to determine whether it is best for a component from a TDMA receiver to be implemented in software, an FPGA, or an application-specific integrated circuit (ASIC) before prototyping the component. The approach chosen uses an activity-based power estimation model and simulation to determine activity in the component. The power model is specialized for each implementation technology using a refinement on a basic, abstract power model. The functional model is specialized similarly, changing the activity estimation based on the implementation technology. The power model and the functional model are composed using a product and a combinator applied to the result to generate a simulation model. The model can then be evaluated to determine the best implementation strategy. A graphical representation of this construction for FPGA, CMOS, and software implementations is shown in Figure 18.4.

The modeling process begins by determining the modeling goals. In the power-aware domain, the modeling process must support making trade-off decisions between implementation technologies. To achieve this, the relationship between the function being performed and how that function consumes power in each implementation technology must be understood. The specification anatomy must lead to a model that examines function and power consumption simultaneously.

To accomplish the trade-off analysis task, functional, power consumption, and power constraint models are required. The interaction between function and power consumption will give a system-level perspective on how power is consumed. The power constraint will indicate what limitations exist on available power. The interaction between power consumption, power constraints, and function will play a role in how the specifications are developed.

For the power-aware modeling activity, the basic Rosetta domain semi-lattice provides appropriate modeling domains for each model type. The power consumption model used is activity-based where power is consumed when a system changes state. The basic model makes no reference to any specific time

297

representation, so the state_based domain is the most appropriate. The functional model is typically given as a discrete time model and best fits in the discrete_time domain. Finally, the power constraint is a simple constant value that cannot be exceeded. The fact that it is constant implies that the static domain is the most appropriate domain.

A discrete event simulation tool that can simulate state_based specifications is also assumed to exist. This is not a part of the standard Rosetta system, but could be provided by an external tool environment. The des domain in the semi-lattice defines requirements for the simulator, allowing functors and combinators to generate models for simulation. It is important to note that one could also explore these specifications using model checking, theorem proving, static analysis, or any number of formal and semi-formal techniques. This is simply a matter of defining a domain and functor for each technique, similar to the des domain.

The relationships between domains in the semi-lattice and the connections between models and domains are depicted in Section 18.2. A segment of the domain semi-lattice is shown with the three models of the component's function, an activity-based power consumption model, and a power constraint model. Also shown is the combinator that generates simulatable models.

18.1   The Basic Models

The first modeling task is writing basic models for power constraints, power consumption, and device function. By keeping the models separate and in their own domains, each model remains focused and reasonably simple. Although some forethought must be given to how they will be composed, for practical purposes they can be written separately.

18.1.1 Power Constraints Model

The power constraint (Figure 18.1) is constant across all time. Thus, the static domain is used to represent the static construct. The facet model simply defines a power variable and asserts that it must be less than the specified limit parameter. The power value is then exported to allow other models to reference it and assert constraints on it. By itself, the powerConstraint model does very little. Remember that it models the *constraint* on power consumption, not power consumption itself. It can easily be instantiated with a specific constraint. Without any other constraints on the exported power value, analysis tools cannot determine if the power constraint is or is not met.

18.1.2 Power Consumption Model

When modeling power consumption using an activity-based model, power is consumed whenever the device changes state. Thus, the power consumption

```
facet powerConstraint(limit::design real)::static is
  export power;
  power::real;
begin
  power <= limit;
end facet powerConstraint;
```

Figure 18.1 Power constraint model for the TDMA component.

```
facet powerConsumption(o::output top;
        leakage,switch::design real)::state_based is
  export power;
  power::real;
begin
  power' = power + leakage + if event(o) then switch else 0 end if;
end facet powerConsumption;
```

Figure 18.2 Power consumption model for the TDMA component.

model (Figure 18.2) is defined in the state_based domain. By choosing the state_based domain rather than the more specific discrete_time domain, we define a common activity-based model that can be specialized for each individual model. The actual value used for state can be refined when the model is actually used. If a discrete_time model were used, the time value would need to be abstracted away if the model were applied in a domain without explicit representation of time.

Unlike the power constraint model, the powerConsumption facet directly describes power consumption in a system. If an event occurs on the output, then the device is assumed to have changed state. In this situation, the device consumes power at a rate specified by the design parameter switch. If not, then only the power specified by leakage is consumed. Total power consumption is accumulated over time to determine the total device power consumption. This is a very naive power model, but it does reflect what is needed — a rough measure of how much power is consumed during the operation of the system.

Before moving forward, let's examine what the product of powerConsumption and powerConstraint models means. The product is simple to specify:

```
limitedPower :: state_based is
    powerConsumption() * state_state_based.gamma(powerConstraint(5e-6))
    sharing {power};
```

The limitedPower model states that both powerConsumption and powerConstraint must be simultaneously satisfied. Alone, each model is satisfied by virtually any legal assignment. What is interesting in the composition is the role of the power variable. Because it is a visible part of both facet states, it is shared in the definitions. In other words, both models reference the same power item. Any value or property asserted on power must satisfy constraints from both facets. If either specification did not export the power item, this interaction would not occur.

In practical terms, the shared power variable means that if the power consumption model causes power to assume a value that is greater than 5e-6, then a contradiction occurs and the engineer knows a problem exists. All that is necessary now is to know when the device changes state.

18.1.3 Functional Model

The TDMA model is a discrete time model that defines functional requirements and takes into account delays through the circuit. Because the device is a time division multiplexer, it is necessary to include an explicit time value to capture the full functionality of the device. Thus, this model is defined in the discrete_time domain to allow this timing information.

The functional model (Figure 18.3) searches for a unique word in a bit stream and passes subsequent bits through until a specified number of bits has been seen. In effect, when the unique ID is detected, one packet of bits is passed through the TDMA receiver.

The specification is not atypical of many discrete_time specifications. When the clock rises, the next state is calculated and the next output is generated. The first three specification blocks determine the next state by watching for a rising edge and then specifying values for variables hit, uniqueID, and bitCounter in the next state. The final specification block observes the state and determines what the next output should be.

Remember that our objective is determining how much power this particular device consumes over time to determine the best implementation choice. However, there is no mention of power or time anywhere in the functional model. This is exactly how it should be. The functional model should reflect the device's function and should not be extended to include power and power constraints. That would defeat the entire purpose of Rosetta and domain-specific modeling. Timing information is present in the model and can be observed by analysis tools. However, using the abstract next function makes moving the model to different domains simpler. The functional model could be written in the state_based domain and moved to the discrete_time domain using a functor.

What will be done to understand the power consumption of this device is to compose the functional model and the power consumption model with the product operator:

```
TDMAPower(i::input real; o::output real)::state_based is
    TDMA(i,o,x"F0F0",1064) * gamma(powerConsumption(i,o,1e-9,2e-8));
```

Like power in the previous product construction, i and o are shared by these models. When o changes based on some input to the TDMA device, the power consumption model will observe that change and will update its power value.

```
facet TDMA(i::input real; o::output real; clk :: in bit
           uniqueID::design word(16);
           pktSize::design natural)::discrete_time is
  uniqueID :: word(16);
  hit :: boolean;
  bitCounter :: natural;
begin

    // Check to see if the unique ID has been seen and whether
    // a full packet has been transmitted.
    hit' =
      risingEdge(clk)
      and ((uniqueID = ID) or hit)
      and not(bitCounter >= pktSize);

    // If not transmitting bits, gather bits for unique ID
    uniqueID' =
      if risingEdge(clk)
        then if hit then x"0000" else sl(uniqueID,...) end if;
        else uniqueID
      end if;

    // If transmitting bits, update the bit counter.
    // Else set the bit counter to 0
    bitCounter' =
      if risingEdge(clk)
        then if hit then bitCounter+1 else 0 end if;
        else bitCounter
      end if;

    // If the unique ID has been seen, output the current bit.
    // Else continue to output the current bit.
    o' =
      if risingEdge(clk)
        then if hit and bitCounter =< pktSize then i else o end if;
        else o
      end if;

end facet TDMA;
```

Figure 18.3 Function model for the TDMA component.

Actually constructing this model takes some additional work, but the basic idea of composing models begins to become more obvious here.

18.2 Composing System Models

One key concept in the case study thus far is that each model is written using a different semantics, independently of other models. Although examples show how the models can be composed, thus far the models are independent and

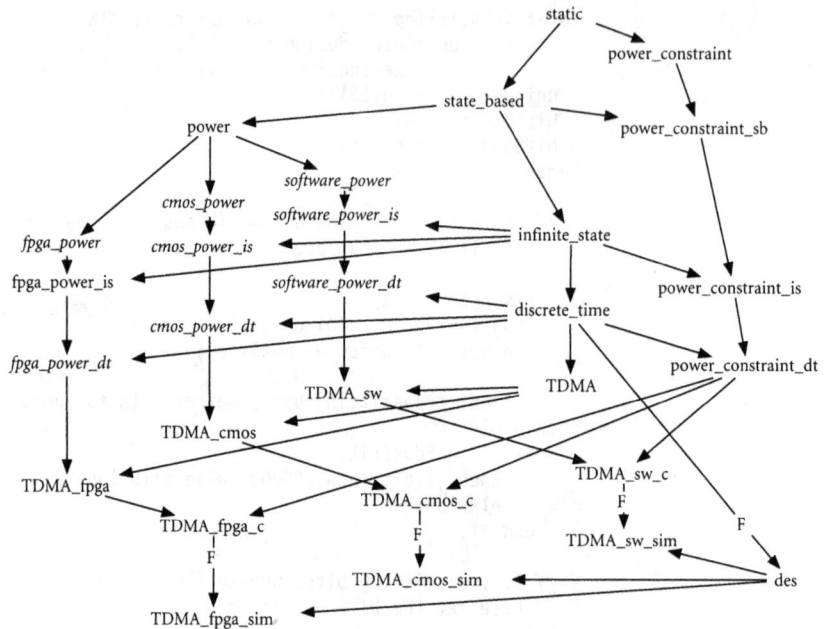

Figure 18.4 Full diagram, showing refinement and composition of device, power, and constraint models.

have no obvious connection other than names of parameters and variables. As shall be seen, the choice of names for items is critical when composing specifications.

The basic models involved in Figure 18.4 have been defined in the previous sections. The remainder of the figure involves defining transformations and compositions of models. Specifically, three things will be accomplished. First, the power constraints model will be refined to the state_based domain for composition with the power consumption model. Then the power consumption model will be specialized three times to reflect the different models of power consumption associated with CMOS, FPGA, and software implementations. Second, the functional model will be abstracted into the state_based domain, again to be composed with the various power consumption models. Finally, a combinator will be applied to create analyzable models from the specification products.

18.2.1 The Composition Approach

The naive approach to composing models is to simply construct a product directly without any model transformations. Unfortunately, this would eliminate all abstractions in our models and make analysis virtually impossible. Specifically, the common domain of these specifications is static, where no concept of time or state change is defined. The result would be a mass of equations that directly encode time and state change. This is not at all desirable.

The solution is choosing a domain where the desired analysis is best performed and constructing a model there. For this analysis, the state_based domain is selected, but other domains could be just as appropriate. The power consumption and power constraint models will be transformed so that they share this less abstract domain with the functional model (Figure 18.5). Although the functional model will lose the concept of time, the most important concept of state change vital to the activity-based power model remains. The specifics of time semantics are not needed.

18.2.2 Refining the Power Model

The process begins with the simplest activity — refining the power constraint model to include state. The refinement of the power constraint model that appears on the right side of Figure 18.4 is shown separately in Figure 18.6. Several functors are composed to construct the morphism that transforms the static power constraint model into a discrete_time model. This collection of transformations is actually quite trivial, as we simply assert that if a property is constant, it must hold at any time step.

In refining the power constraint model we are relying on a default interaction defined between the static and state_based domains. This interaction defines the functor static_state_based.gamma that transforms a specification from the static domain into a specification in the state_based domain. This functor and

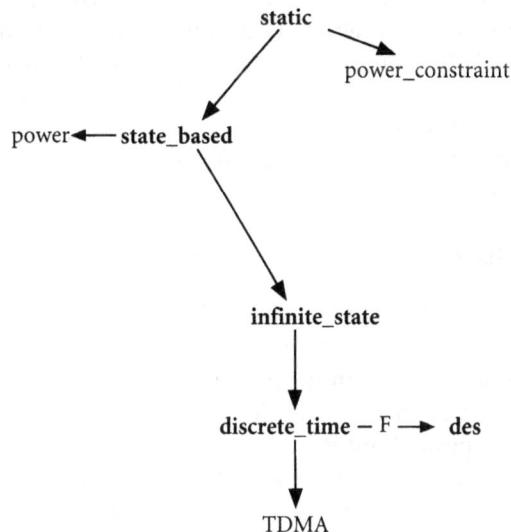

Figure 18.5 Refinement of domains to a functional model.

Figure 18.6 diagram:

```
                    static
                   /      \
                  /        → power_constraint
                 ↓                    |
          state_based ───→            ↓
             |            power_constraint_sb
             |                         |
             ↓                         |
        infinite_state ──→             |
             |          power_constraint_is
             ↓                  \      |
        discrete_time            \     ↓
          ╱ - - ─ ┐   ↓      → power_constraint_dt
         ╱         ↓  TDMA         /
   TDMA_sw ←───────            ╱
        \                    ╱
         \                 ╱
          →  TDMA_sw_c  ←
```

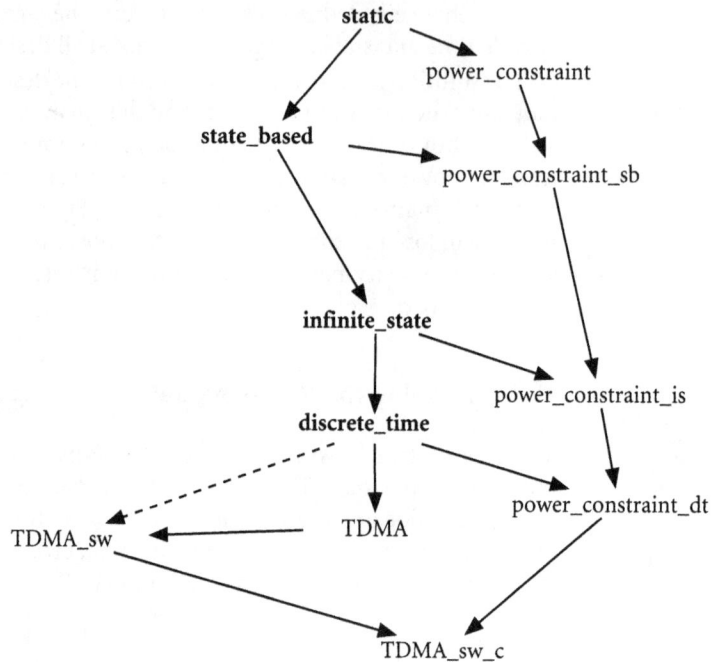

Figure 18.6 Refinement of the power constraint to a state-based model.

functors like it, i.e., that move down the domain semi-lattice, are simple to construct by simply constructing the extension that generates the more concrete domain from the more abstract domain. Whenever one domain is below another in the semi-lattice, the lower domain is created by extending the upper domain. In Chapter 15 these functors are given the name gamma and are defined between each pair of related domains.

In this case the functor used is defined in the interaction static_state_based and has the following signature:

```
gammaS[State::type]() from x::static to state_based(State);

gamma() from x::static to state_based;
```

There are two forms of this functor. Both assert that in the new specification, all terms from the old specification are true in every state. Thus, power must be less than the power limit in every state. The parameter in gammaS is used to provide a set of states to define the state type. In this case, a universally quantified parameter is used to allow the type to be inferred. If the state is held abstract, an identical functor, gamma, is available without the parameter. In this case, the state is unknown. Thus, gamma is initially chosen.

The functor can be used by including the following declaration in the specification:

```
powerConstraint_sb(x::design real) :: state_based is
    static_state_based.gamma(powerConstraint(x));
```

A new facet now exists called `powerConstraint_sb`, parameterized over the power limit. This facet can now be used like any other. Of course, the declaration is not required. The functor may be used directly as a term anywhere in the specification where a facet of type `state_based` is expected.

18.2.3 Transforming the Power Consumption Model

With a state-based model of the power constraint constructed, the technology-specific power consumption models can now be constructed. To achieve this, the original power consumption model is transformed to represent power consumption in multiple different implementation technologies. The "refinement" performed here is achieved simply by instantiating parameters in the power consumption model with appropriate values. Of course, a more sophisticated transformation could be performed when more detail is known. However, an informed implementation decision can be made even in the presence of incomplete information using this model.

The technology-specific power consumption models are defined as follows:

```
CMOSpowerConsumption(i::input real; o::output real) :: state_based
    is powerConsumption(i,o,1e-9,2e-8);

FPGApowerConsumption(i::input real; o::output real) :: state_based
    is powerConsumption(i,o,1e-9,2e-8);

SWpowerConsumption(i::input real; o::output real) :: state_based
    is powerConsumption(i,o,1e-9,2e-8);
```

One does not typically look at parameter instantiation as refinement, but that is exactly what it is in this case. Instantiating the design parameters is literally the equivalent of substituting values for variables. Thus, the resulting specifications are in fact refinements of the original. Both the input and output parameters are retained to allow the facets to be composed structurally and combined with other specifications.

Now the new technology-specific power models can be composed with the power constraint. All exist in the `state_based` domain, so there is no need to transform specifications to other domains. The new facet models are easily defined as:

```
CMOSpowerLimited(i::input real o::output real) :: state_based;
    is CMOSpowerConsumption(i,o) * powerConstraint_sb(5e-6);

FPGApowerLimited(i::input real o::output real) :: state_based;
    is FPGApowerConsumption(i,o) * powerConstraint_sb(5e-6);

SWpowerLimited(i::input real o::output real) :: state_based
    is SWpowerConsumption(i,o) * powerConstraint_sb(5e-6);
```

These specifications all share a similar form and construct a product of a technology-specific power consumption model and a power constraint. Because the power constraint and the power consumption model export the power item, this value is considered shared between the specifications, and any constraints on it must be mutually consistent. This is precisely what is desired — a power consumption model limited by a specified power constraint. When we drive the values of i and o, the power consumption model will increase consumed power and the constraint model will compare with its constraint value.

What remains now is composition with the functional model. There are two equally valid approaches. The first refines the state_based power consumption models into the discrete_time domain and constructs a product with the functional model. The second abstracts the functional model to the state_based domain and constructs the product there. Both approaches are examined in the following sections.

18.2.4 Refining the Power Models

The refinement of constrained power consumption models into the discrete_ time domain is shown in Figure 18.7. The figure shows one of three constructions that generate the FPGA, CMOS, and software power consumption models on the left side of Figure 18.4. The approach mimics the approach used to move the power constraint into the state_based domain. Specifically, a functor is applied to the constrained power models that moves them from state_based to discrete_time.

In this case, the functor is actually the composition of two functors for moving from state_based to infinite_state and infinite_state to discrete_time. Defining the new functor is simply defining a function whose domain and range are facet types. Specifically:

```
gammaSBDT(f::state_based)::discrete_time is
    state_based_infinite_state.gamma.infinite_state_discrete_time.gamma
```

Functors in the built-in interaction libraries accomplish the refinement between individual domains. Getting the power models to the discrete_time domain is now a matter of applying the new functor:

```
let gammaSBDT(f::state_based)::discrete_time be
    state_based_infinite_state.gamma.infinite_state_discrete_time.gamma
  in
    CMOSpowerLimit_dt(i::input real; o::output real) :: discrete_time is
      gammaSBDT(CMOSpowerLimit(i,o));

    FPGApowerLimit_dt(i::input real; o::output real) :: discrete_time is
      gammaSBDT(FPGApowerLimit(i,o));

    SWpowerLimit_dt(i::input real; o::output real) :: discrete_time is
      gammaSBDT(SWpowerLimit(i,o));
end let;
```

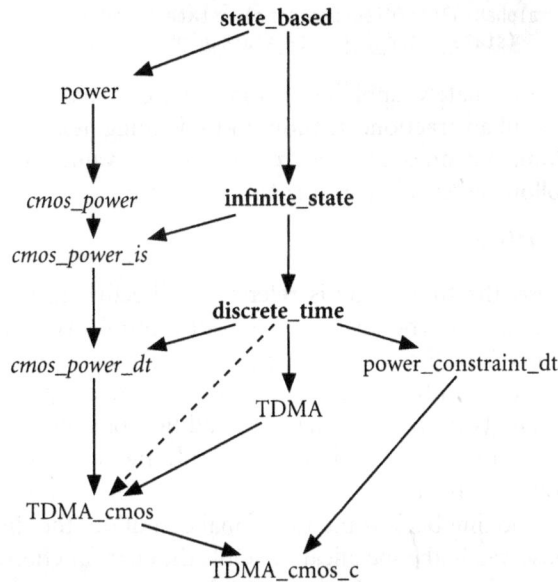

Figure 18.7 Refinement of a basic state-based power model to a CMOS power consumption model.

The product can now be directly formed with the functional model and a discrete_time specification that accounts for power consumption results:

```
CMOS_TDMA(i::input real; o::output real) :: discrete_time is
    CMOSpowerLimit_dt(i,o) *
    TDMA(i,o,x"F0F0",1064);
FPGA_TDMA(i::input real; o::output real) :: discrete_time is
    FPGApowerLimit_dt(i::input real; o:output real) *
    TDMA(i,o,x"F0F0",1064);
SW_TDMA(i::input real; o::output real) :: discrete_time is
    SWpowerLimit_dt(i::input real; o:output real) *
    TDMA(i,o,x"F0F0",1064);
```

18.2.5 Abstracting the Functional Model

The alternative is moving the functional model to state_based and performing the composition there. The advantage is that the resulting model is simpler and more abstract. What we want here is the opposite of refinement — abstraction of the discrete_time model to the state_based domain. This is done using the alpha functors defined in the state_based_infinite_state and infinite_state_discrete_time interactions. Specifically, the alpha instance needed here is:

```
alphaDTSB(f::discrete_time)::state_based is
  (state_based_infinite_state.alpha.infinite_state_discrete_time.alpha)
```

Unfortunately, applying this instance of `alpha` may not always result in a successful abstraction. In addition to defining `next` concretely, the `discrete_time` domain defines time concretely as natural value. Thus, it is quite legal to have the following term in a `discrete_time` facet:

```
v@(t+5) = v+1;
```

Here, the time value is referenced directly rather than indirectly using `next`. Making its type abstract may cause problems if the addition operator is not defined on that abstract type. All is not lost. Remember that the `state_based` domain can be parameterized over the state type, making it possible to specify a state type that does have an addition operator. However, this does not truly result in an abstraction, but simply moving a complete definition to a new frame of reference.

Looking back at the functional definition, the time value, `t`, is never directly observed in the specification. Only the `next` function is used to specify movement from state to state. In this case it is possible to simply replace the `discrete_time` domain with the `state_based` domain. Thus, the functor from the built-in interactions can safely perform this operation.

With the abstract functional model in the `state_based` domain along with the power consumption model, the product can be formed in the `state_based` domain or a combinator can be used to generate a new model. This process is identical to that used when the power consumption models are made concrete, and thus will not be repeated here. The point of emphasis here is that if abstractions like this are to be used, care must be taken when writing specifications. When the intent is to abstract a model to a new domain, writing the model using vocabulary from the abstract domain has advantages. If this cannot be done, then writing a new `alpha` specifically for the situation is the most appropriate approach.

18.3 Constructing the Simulations

Following the transformation of the constraint models into the `discrete_time` domain where the functional model exists, products are used to construct a systems model. The functional model is used to generate activity information for the FPGA, CMOS, and software power consumption models while the power constraint model simply asserts a condition that must hold continuously in each model.

In both cases, an algebra combinator is used to construct a composite model from the product models. Details of the combinator are specific to the tool set used for performing the analysis. However, the interaction necessary for the composition and transformation has the form:

```
interaction discrete_time_des() between discrete_time and des is
begin
   begin translators;
   end translators;
   begin functors
      alpha() from des to discrete_time;
      gamma() from discrete_time to des;
   end functors;
   begin combinators
      alphaC() from x::des and y::des to discrete_time is
         alpha(x) * alpha(y);
      gammaC() from x::discrete_time and y::discrete_time to des is
         gamma(x) * gamma(y);
   end combinators;
end interaction discrete_time_des;
```

The necessary algebra combinator's signature is is defined in the **combinators** section and has the signature:

```
gammaC(f::discrete_time,g::discrete_time)::des;
```

The final analysis models are:

```
CMOSsim(i::input real; o::output real)::des is
   gammaC(CMOSpowerLimit_dt(i,o),TDMA(i,o,x"F0F0",1064));

FPGAsim(i::input real; o::output real)::des is
   gammaC(FPGApowerLimit_dt(i,o),TDMA(i,o,x"F0F0",1064));

SWsim(i::input real; o::output real)::des is
   gammaC(SWpowerLimit_dt(i,o),TDMA(i,o,x"F0F0",1064));
```

Figure 18.8 shows the combinator applied to construct simulation models from the product models. Like functors, combinators move models from one domain to another. The distinction is that combinators operate on pairs of models to generate a single model. Where products and sums cannot model the effects of interactions between domains, combinators can, because they generate new models.

18.4 Wrap Up

This power-aware design example is among the oldest Rosetta specifications and has been used as an analysis example through many language revisions and tool instances. The structure of the specification is classic Rosetta. A collection of models is defined and composed to support cross-domain analysis. Although tool-specific analysis results are not germane to this presentation, we can speak to many of the emergent issues and common analysis results.

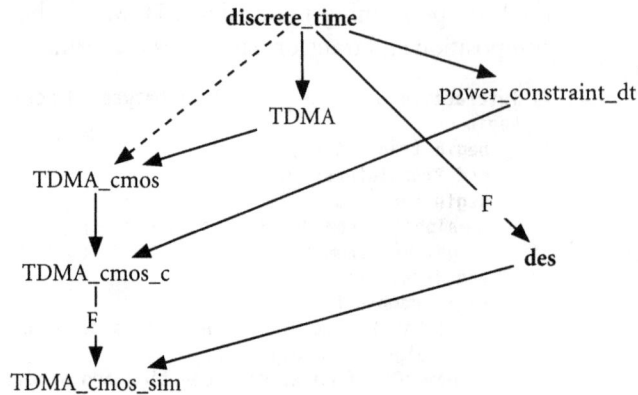

Figure 18.8 A combinator used to generate simulation models from discrete time models.

18.4.1 Analysis Results

Now that we have the models, what do we do with them? These models have been analyzed in one form or another in Matlab, a custom Java analysis environment, and in the specialized Raskell Rosetta analysis environment. Across the board, analysis results depend more on the quality of data provided to models written, than on the tool or models used. However, we were able to learn things about interacting specifications in all tools that were beyond what simple simulation would show us.

Analysis results reveal that the software solution was the least power efficient. Our goal was demonstrating Rosetta modeling, not developing accurate power models, meaning the actual results should not be taken too seriously. However, we were able to observe changes in the power profile of the system with respect to changes in the execution profile. Ultimately this was our objective — understanding the impacts of local design decisions on global system-level properties. With more accurate models and a more realistic process, there is little question that analysis results would provide useful information.

In the most detailed analysis activity using Raskell, we were able to use an algebra combinator to move results from a functional analysis to the power analysis domain. This allowed us to estimate the power utilization based on the actual function being performed. The largest impact observed was in estimating the power cost of the software implementation. Because we were modeling software running on a CPU and the impacts of software state change on hardware state change, estimates were better than simply estimating power consumption on the CPU in general. However, this gain is not without cost. The actual models for power consumption are far more difficult to generate than are models for hardware implementations. This is a reflection of the intellectual distance between the software models and hardware realizations that actually consume power.

```
package powerAwareComponentModels() is
  export all;
  facet powerConstraint(limit::design real)::static is
    export power;
    power::real;
  begin
    power <= limit;
  end facet powerConstraint;

  facet powerConsumption(o::output top;
          leakage,switch::design real)::state_based is
    export power;
    power::real;
  begin
    power' = power + leakage + if event(o) then switch else 0 end if;
  end facet powerConsumption;

  facet TDMA(i::input real; o::output real; clk :: in bit
          uniqueID::design word(16);
          pktSize::design natural)::discrete_time is
    uniqueID :: word(16);
    hit :: boolean;
    bitCounter :: natural;
  begin

    hit' = risingEdge(clk)
      and ((uniqueID = ID) or hit)
      and not(bitCounter >= pktSize);

    uniqueID' = if risingEdge(clk)
          then if hit then x"0000" else sl(uniqueID,...) end if;
          else uniqueID
      end if;

    bitCounter' = if risingEdge(clk)
          then if hit then bitCounter+1 else 0 end if;
          else bitCounter
        end if;

    o' = if risingEdge(clk)
          then if hit and bitCounter =< pktSize then i else o end if;
          else o
        end if;
  end facet TDMA;
end package powerAwareComponentModels;
```

Figure 18.9 Component models.

In addition to simulation analysis, formal techniques were used to perform some type analysis and some functional analysis. We used a theorem prover for these assessments. For power analysis, particularly using activity-based models, simulation is the best approach even though it is only semi-formal. Do not underestimate the value of formal semantics in simulation. Because the specification has a precise meaning, the correctness of simulation tools can be addressed.

In this example, simulators were synthesized using formal techniques that guarantee their correctness. Such synthesis is not possible without Rosetta's formal underpinnings.

18.4.2 Modeling Overview

Looking back at Figure 18.4, it should now be clear how the diagram is formed. Although morphisms in the diagram create some clutter, the shapes of different activities emerge, now that each activity has been identified. The original constraint models were refined as necessary to generate state_based or discrete_time models. These models were composed and models were generated using an algebra combinator. The resulting simulation models in the discrete event (des) domain were simulated and analyzed. Although Figure 18.4 is busy, it does represent the constructions necessary to construct the simulation models.

The models are presented as a complete system in Figures 18.9 through 18.11. The final power-aware case study model is split up between several packages for presentation. Figure 18.9 shows a package containing the several component models. Figure 18.10 uses the components package and defines power consumption models for each component. Finally, Figure 18.11 imports the packages containing component and power models, composing the models to define the final analysis models.

```
use powerAwareComponentModels;
package powerAwarePowerModels() is
  export all;

  powerConstraint_sb(x::design real) :: state_based is
    static_state_based.gamma(powerConstraint(x));

  CMOSpowerConsumption(i::input real; o::output real) :: state_based
    is powerConsumption(i,o,1e-9,2e-8);

  FPGApowerConsumption(i::input real; o::output real) :: state_based
    is powerConsumption(i,o,1e-9,2e-8);

  SWpowerConsumption(i::input real; o::output real) :: state_based
    is powerConsumption(i,o,1e-9,2e-8);

  CMOSpowerLimited(i::input real; o::output real) :: state_based
    is COMSpowerConsumption(i,o) * powerConstraint_sb(5e-6);

  FPGApowerLimited(i::input real; o::output real) :: state_based
    is powerConsumption(i,o) * powerConstraint_sb(5e-6);

  SWpowerLimited(i::input real; o::output real) :: state_based
    is powerConsumption(i,o) * powerConstraint_sb(5e-6);

end package powerAwarePowerModels;
```

Figure 18.10 Power consumption models.

```
use powerAwareComponentModels;
use powerAwarePowerModels;
package powerAwareCaseStudy() is
  export all;

  sb2dt(f::state_based)::discrete_time is
    infinite_state_discrete_time.gamma.state_based_infinite_state.gamma;

  CMOSpowerLimit_dt(i::input real; o::output real) :: discrete_time is
    sb2dt(CMOSpowerLimit(i,o));

  FPGApowerLimit_dt(i::input real; o::output real) :: discrete_time is
    sb2dt(FPGApowerLimit(i,o));

  SWpowerLimit_dt(i::input real; o::output real) :: discrete_time is
    sb2dt(SWpowerLimit(i,o));

  CMOSsim(i::input real; o::output real)::des is
    gammaC(CMOSpowerLimit_dt(i,o),TDMA(i,o,x"F0F0",1064));

  FPGAsim(i::input real; o::output real)::des is
    gammaC(FPGApowerLimit_dt(i,o),TDMA(i,o,x"F0F0",1064));

  SWsim(i::input real; o::output real)::des is
    gammaC(SWpowerLimit_dt(i,o),TDMA(i,o,x"F0F0",1064));

end package powerAwareCaseStudy;
```

Figure 18.11 Final model for the power-aware case study.

Power-Aware Modeling Revisited

Although the power-aware model for the TDMA system developed in Chapter 18 supports some predictive analysis in the absence of virtually all design detail, the power model is quite naive and the values discovered will not remain accurate as design decisions are made. The power-aware model is revisited here in several different ways to explore new analysis possibilities.

Technology-specific functional models allow representation of the TDMA algorithm in a manner most appropriate for a specific technology. Although the function of the TDMA remains the same, moving from hardware to software certainly changes the algorithm and implementation. New models for TDMA functions represent technology-specific implementations for functions in a manner similar to that for technology-specific power consumption models.

Using decomposition to represent the TDMA model structurally adds design detail and allows refinement of the power consumption model. By modeling power consumption in individual components, power consumption is more precisely modeled. Using facets to structurally model constraints as well as functions results in clean, focused structural specifications.

Finally, combining decomposition with technology-specific functional and power models adds yet more detail to the power model. Implementation technology for each component is included, as well as system architecture. Again, structural modeling of constraints as well as function supports highly decoupled models.

19.1 Technology-Specific Functional Models

The first technique explores the use of different functional models for different technologies. In the original model, the CMOS, FPGA, and software models all used the same functional model. At the requirements level, this is appropriate. However, as the design is refined it becomes quite apparent that the functional implementations differ from technology to technology. Thus, power consumption will vary in ways that cannot be accounted for in the original model.

To construct this new model, three new functional models must be developed. These new models need not be written in the same domain as the original model, but care must be taken to use domains that can be linked back to analysis tools. For the purposes of this example, CMOS, FPGA, and software models are developed in the discrete_time domain. Following are interfaces for these models:

```
facet interface CMOS_TDMA(i::input real; o::output real)::discrete_time is
end facet interface CMOS_TDMA;

facet interface FPGA_TDMA(i::input real; o::output real)::discrete_time is
end facet interface FPGA_TDMA;

facet interface SW_TDMA(i::input real; o::output real)::discrete_time is
end facet interface SW_TDMA;
```

These models are composed with the technology-specific power models using combinators in the same manner used in Chapter 18. When the combinator generating simulation models is applied to these models, the result is an analysis model with more detail, due to the inclusion of technology-specific information.

```
CMOS_TDMA(i::input real; o::output real) :: discrete_time is
 CMOSpowerLimit_dt(i::input real; o:output real) *
 CMOS_TDMA(i::input real;o::output real);

FPGA_TDMA(i::input real; o::output real) :: discrete_time is
 FPGApowerLimit_dt(i::input real; o:output real) *
 FPGA_TDMA(i::input real;o::output real);

SW_TDMA(i::input real; o::output real) :: discrete_time is
 SWpowerLimit_dt(i::input real; o:output real) *
 SW_TDMA(i::input real;o::output real);
```

19.2 Configurable Components

An interesting variant on the technology-specific model uses a **case** expression to allow a single, parameterized model to be configured as any of the technology-specific models. Using a constructed type to indicate the technology type and **case** expression to choose the model is the approach taken here.

First, the technology data type is defined, allowing selection of the technology type. The data type defines an enumerated type with three values associated with one technology:

```
technology :: type is data
  cmos :: cmosp |
  fpga :: fpgap |
  sw :: swp
end data;
```

Remember that this data type must be visible outside the new TDMA facet. Thus, it cannot be defined locally to the TDMA facet. Using a package to define a module including both definitions is the most appropriate approach.

Next, the model is defined by adding a new parameter, t, of type technology. This parameter of kind **design** allows configuration of the component and serves as the selection variable in the **case** expression:

```
TDMA(t::design technology i::input real; o::output real;
    clk :: in bit; uniqueID::design word(16);
    pktSize::design natural) :: discrete_time is
case t of
  {cmos()} -> CMOS_TDMA(i,o,clk,uniqueID,pktSize) |
  {fpga()} -> FPGA_TDMA(i,o,clk,uniqueID,pktSize) |
  {sw()} -> SW_TDMA(i,o,clk,uniqueID,pktSize)
end case;
```

The technology selection parameter is of kind **design** because it cannot change during evaluation. If a reason existed to change technologies at run-time, the parameter restriction could be removed. Such a model would be useful for analyzing the process of moving an operational system from one technology to another to understand run-time risks.

This new TDMA model can be included in a design and the technology type parameter can be used to indicate the type of component to include. The value of this approach is being able to quickly reconfigure a model to check properties for different technology types. For example, the following term represents the FPGA implementation of the TDMA component:

```
tdma_comp: TDMA(fpga,i,o,clk,x"F0F0",1064);
```

This definition assumes that i, o, and clk are items visible in its scope.

19.3 Decomposition

The second technique explores decomposing the functional model into a structural model. This technique increases accuracy by looking at finer grained state changes and thus finer grained power consumption. Figure 19.1 shows an abstract block diagram of the TDMA receiver. Six interconnected components are defined, driven by a single clock. As the TDMA receiver is a single processing block, its architecture is largely dataflow in nature. Data flowing into and out of components is complex, representing the in phase and quadrature components of the signal.

The definitions in Figure 19.2 are for the individual components comprising the structural block diagram from Figure 19.1 defining interfaces and functional elements. These definitions are no different than the functional models written for the original model. What is happening is that the functional definition is being pushed deeper into the design representation. This will always result when making design decisions.

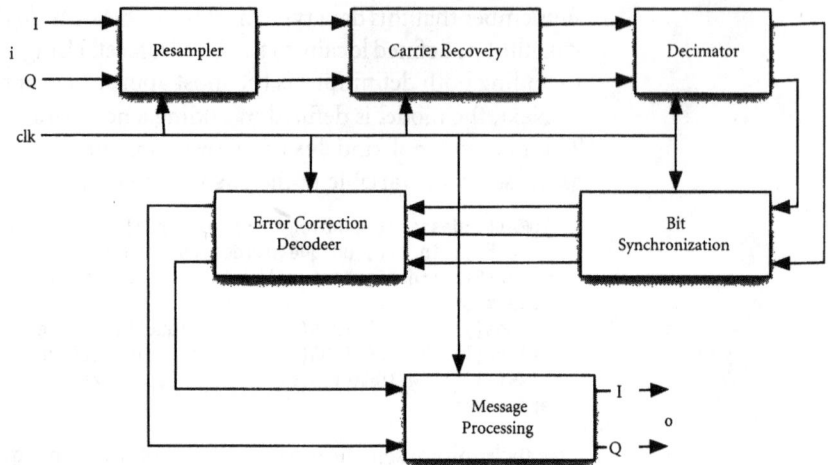

Figure 19.1 TDMA block diagram.

```
facet interface resampler
  (i::input complex; o::output complex; clk::input bit)::discrete_time is
end facet interface resampler;

facet interface carrierRecovery
  (i::input complex; o::output complex; clk::input bit)::discrete_time is
end facet interface carrierRecovery;

facet interface decimator
  (i::input complex; o::output complex ,clk::input bit)::discrete_time is
end facet interface decimator;

facet interface bitSynchronization
  (i::input complex; o::output complex ,clk::input bit)::discrete_time is
facet interface bitSynchronization;

facet interface errorCorrection
  (i::input complex; o::output complex ,clk::input bit)::discrete_time is
end facet interface errorCorrection;

facet interface messageProcessor
  (i::input complex; o::output complex ,clk::input bit)::discrete_time is
end facet interface messageProcessor;
```

Figure 19.2 Model interfaces for TDMA components.

The structural design in Figure 19.3 accomplishes three modeling tasks. First, each component model is composed with the CMOS power consumption model and instantiated within the component. The result is a collection of specification products representing each component's function and power consumption

```
facet TDMAstruct
  (i::input complex; o::output complex; clk::input bit)::discrete_time is
    export power;
    power :: real;
    ec2mp,bs2ec,d2bs,cr2d,r2cr :: complex;
    gammaSBDT(f:state_based)::discrete_time is
      (state_based_infinite_state.gamma . infinite_state_discrete_time.gamma);

begin
  c1: resampler(i,r2cr,clk)
      * gammaSBDT(CMOSpowerConsumption(i,r2cr));
  c2: carrierRecovery(r2cr,cr2d,clk)
      * gammaSBDT(CMOSpowerConsumption(r2cr,cr2d));
  c3: decimator(cr2d,d2bs,clk)
      * gammaSBDT(CMOSpowerConsumption(cr2d,d2bs));
  c4: bitSynchronization(d2bs,bs2ec,clk)
      * gammaSBDT(CMOSpowerConsumption(d2bs,bs2ec));
  c5: errorCorrection(bs2ec,ec2mp,clk)
      * gammaSBDT(CMOSpowerConsumption(bs2ec,ec2mp));
  c6: messageProcessor(ec2mp,o,clk)
      * gammaSBDT(CMOSpowerConsumption(ec2mp,o));
  power' = power + c1.power + c2.power + c3.power
           + c4.power + c5.power + c6.power;
end facet TDMAstruct;
```

Figure 19.3 TDMA with power calculation.

model. Next, interconnections are made between individual components and from components to the TDMA receiver interface. Finally, the power associated with the TDMA receiver is calculated by summing the powers from the individual components.

The result of this construction is somewhat different than earlier models in that it does not include the power constraint. The power value exported from the model is the consumed power, not a power limit. There are two ways to include the power constraint. One is to simply compose the TDMA model with the power constraint at the component level. This is achieved using the state_based power constraint model and constructing the product with the structural TDMA after moving the state_based model to discrete_time:

```
TDMA(i::input complex; o::output complex; clk::input bit)::discrete_time is
  TDMAstruct(i,o,clk) *
  state_based_infinite_state.gamma(
    infinite_state_discrete_time.gamma(powerConstraint_sb(i,o)));
```

An alternative approach is to budget power across components and add the power constraint model to each instantiated component. The distinction is that when composing the structural model with the the system power constraint, analysis will only reveal a problem at the systems level. It will not indicate the component or components causing the problem. By budgeting the power constraint across components, analysis will reveal problems at the component

```
facet TDMAstruct
  (i::input complex; o::output complex; clk::input bit)::discrete_time is
    export power;
    power :: real;
    ec2mp,bs2ec,d2bs,cr2d,r2cr :: complex;
    gammaSBDT(f:state_based)::discrete_time is
      (state_based_infinite_state.gamma . infinite_state_discrete_time.gamma);
begin
  c1: resampler(i,r2cr,clk)
        * (gammaSBDT(CMOSpowerConsumption(i,r2cr)
          * static_state_based.gamma(powerConstraint(1e-6))));
  c2: carrierRecovery(r2cr,cr2d,clk)
        * (gammaSBDT(CMOSpowerConsumption(r2cr,cr2d)
          * static_state_based.gamma(powerConstraint(2e-6))));
  c3: decimator(cr2d,d2bs,clk)
        * (gammaSBDT(CMOSpowerConsumption(cr2d,d2bs)
          * static_state_based.gamma(powerConstraint(1e-6))));
  c4: bitSynchronization(d2bs,bs2ec,clk)
        * (gammaSBDT(CMOSpowerConsumption(d2bs,bs2ec)
          * static_state_based.gamma(powerConstraint(3e-6))));
  c5: errorCorrection(bs2ec,ec2mp,clk)
        * (gammaSBDT(CMOSpowerConsumption(bs2ec,ec2mp)
          * static_state_based.gamma(powerConstraint(1e-6))));
  c6: messageProcessor(ec2mp,o,clk)
        * (gammaSBDT(CMOSpowerConsumption(ec2mp,o)
          * static_state_based.gamma(powerConstraint(1e-6))));
end facet TDMAstruct;
```

Figure 19.4 Structural TDMA receiver model with power budgeted to individual components.

```
facet TDMAfunction
  (i::input complex; o::output complex; clk::input bit)::discrete_time is
    export power;
    power :: real;
    ec2mp,bs2ec,d2bs,cr2d,r2cr :: complex;
begin
  c1: resampler(i,r2cr,clk);
  c2: carrierRecovery(r2cr,cr2d,clk);
  c3: decimator(cr2d,d2bs,clk);
  c4: bitSynchronization(d2bs,bs2ec,clk);
  c5: errorCorrection(bs2ec,ec2mp,clk);
  c6: messageProcessor(ec2mp,o,clk);
end facet TDMAfunction;
```

Figure 19.5 Structural TDMA receiver model.

level. Unfortunately, the model from Figure 19.4 is unwieldy and defeats the separation-of-concerns goals that Rosetta promotes. An alternative mechanism exists to form the product model using a products of structural facets.

Figures 19.5 through 19.7 structurally model function, power constraints, and power consumption for the TDMA receiver model. These models are separate in the same sense that the original TDMA models were separate. Using the product, they are easily composed in the same manner as the individual models:

```
facet TDMApowerBudget()::discrete_time is
    export power;
    power :: real;
begin
  c1: static_state_based.gamma(powerConstraint(1e-6)));
  c2: static_state_based.gamma(powerConstraint(2e-6)));
  c3: static_state_based.gamma(powerConstraint(1e-6)));
  c4: static_state_based.gamma(powerConstraint(3e-6)));
  c5: static_state_based.gamma(powerConstraint(1e-6)));
  c6: static_state_based.gamma(powerConstraint(1e-6)));
  power = c1.power + c2.power + c3.power
        + c4.power + c5.power + c6.power;
end facet TDMApowerBudget;
```

Figure 19.6 Structural model of power budgets for TDMA components.

```
facet TDMAconsumption
  (i::input complex; o::output complex)::discrete_time is
    export power;
    power :: real;
    ec2mp,bs2ec,d2bs,cr2d,r2cr :: complex;
    gammaSBDT(f:state_based)::discrete_time is
      (state_based_infinite_state.gamma . infinite_state_discrete_time.gamma);
begin
  c1: gammaSBDT(CMOSpowerConsumption(i,r2cr));
  c2: gammaSBDT(CMOSpowerConsumption(r2cr,cr2d));
  c3: gammaSBDT(CMOSpowerConsumption(cr2d,d2bs));
  c4: gammaSBDT(CMOSpowerConsumption(d2bs,bs2ec));
  c5: gammaSBDT(CMOSpowerConsumption(bs2ec,ec2mp));
  c6: gammaSBDT(CMOSpowerConsumption(ec2mp,o));
  power' = power +  c1.power + c2.power + c3.power
        + c4.power + c5.power + c6.power;
end facet TDMAconsumption;
```

Figure 19.7 Structural model of power consumption for TDMA components.

```
TDMA(i::input complex; o::output complex; clk::input bit)::discrete_time is
  TDMAfunction(i,o,clk)
    * gammaSBDT(TDMAconsumption(i,o,clk)
      * static_state_based.gamma(TDMApowerBudget()));
```

This model is semantically identical to the previous model. The product operation is applied to each of the terms comprising the original structural models. Term labels are used to determine how the operations are applied. The product operations are distributed across like-named terms. For example, the c1 term resulting from the product is defined as:

```
TDMAfunction.c1
* gammaSBDT(TDMAconsumption.c1
        * static_state_based.gamma(TDMApowerBudget.c1));
```

Other terms are handled similarly, resulting in exactly the model defined earlier.

19.4 Mixed Technology Systems

The final model combines the use of separate technology models and structural modeling. Figures 19.8 and 19.9 show functional and power decomposition of the receiver, respectively. Note the use of different models for different implementation technologies. Formation of the product model is identical to that in the previous example and results in a yet more accurate model.

```
facet TDMAconsumption
  (i::input complex; o::output complex)::discrete_time is
    export power;
    power :: real;
    ec2mp,bs2ec,d2bs,cr2d,r2cr :: complex;
    gammaSBDT(f:state_based)::discrete_time is
      (state_based_infinite_state.gamma . infinite_state_discrete_time.gamma);

  begin
  c1: gammaSBDT(FPGApowerConsumption(i,r2cr));
  c2: gammaSBDT(CMOSpowerConsumption(r2cr,cr2d));
  c3: gammaSBDT(SWpowerConsumption(cr2d,d2bs));
  c4: gammaSBDT(FPGApowerConsumption(d2bs,bs2ec));
  c5: gammaSBDT(SWpowerConsumption(bs2ec,ec2mp));
  c6: gammaSBDT(SWpowerConsumption(ec2mp,o));
  power' = power +  c1.power + c2.power + c3.power
          + c4.power + c5.power + c6.power;
  end facet TDMAconsumption;
```

Figure 19.8 Structural model of power consumption for TDMA components using mixed implementation technology.

Mixed technology models are particularly useful when performing hardware/software co-design. Such models may be automatically produced by synthesis tools and analysis results may be used to select the best implementation approach. Synthesis combinators also take advantage of these models by applying technology-specific synthesis techniques to individual functional components.

19.5 Wrap Up

The objective of this case study is to explore alternatives to the initial power-aware case study and discuss refinement of system-level models. In this chapter, heterogeneous design is truly explored for the first time by allowing for different functional models for components using different implementation technologies. Additionally, the option of implementing different system functions simultaneously using different implementation technologies can be explored.

```
facet TDMAfunction
  (i::input complex; o::output complex; clk::input bit)::discrete_time is
    export power;
    power :: real;
    ec2mp,bs2ec,d2bs,cr2d,r2cr :: complex;
begin
  c1: FPGAresampler(i,r2cr,clk);
  c2: CMOScarrierRecovery(r2cr,cr2d,clk);
  c3: SWdecimator(cr2d,d2bs,clk);
  c4: FPGAbitSynchronization(d2bs,bs2ec,clk);
  c5: SWerrorCorrection(bs2ec,ec2mp,clk);
  c6: SWmessageProcessor(ec2mp,o,clk);
end facet TDMAfunction;
```

Figure 19.9 Structural TDMA receiver model using mixed technology components.

The introduction of technology-specific functional models has the greatest potential for increased modeling detail. As noted in Chapter 18, software uses a different underlying computational model and changes states in very different ways, as compared to hardware implementations of the same function. Technology-specific function models make it possible to define and analyze such systems. Using configurable models to select from among alternative technologies makes it simpler to configure and analyze models.

Structural decomposition is an important refinement step in any system design process. As design detail is added to functional models, it also propagates to constraint models. In effect, the constraint models follow design decisions, allowing more detailed and precise analysis. This evolution is vital as requirements flow through the design process or performance requirements become stale artifacts of initial design decisions.

When system components can be implemented using different technologies, we begin to explore the benefits of heterogeneity. Allowing models for components to be configurable over several implementation fabrics dramatically improves understanding of trade-offs among performance requirements and implementation fabrics. Using simulation, the system impacts of design fabrics and their interactions can be explored. If significant resources are brought to bear, one can imagine finding optimum or near-optimum implementations.

System-Level Networking

The system-level network case study models the behavior of changing network infrastructure with respect to a constant set of access control requirements. The example is motivated by the problem of moving a portable computer from one network to another, seamlessly maintaining functionality without violating security constraints. Figure 20.1 illustrates the problem where in a portable computer is carried between an office environment that is closed and secure, a home environment that is somewhat controlled, and a completely open environment such as a coffee shop, where the environment is not known or controllable. The user of the portable computer would like to work on the same task while satisfying security requirements. In effect, a trade-off between accessibility and access control is continually being made.

This case study differs substantially from the RTL and power-aware studies in that the system implementation is not a part of the modeling process. Certainly architectural issues are addressed, but the biggest issue is developing a reasonable requirements model. Thus, we will not develop models of firewalls and routers, but models of working environments and resource access restrictions.

Although models may differ, the methodology followed by the networking case study is virtually identical to that in the previous case studies. First, decide what information must be known. Second, select domains for the basic models that are as abstract as possible. Then define basic models for components and compose models to define systems. Finally, generate analysis models from composite models by moving them to different domains.

20.1 The Basic Models

There are three types of models involved in this example, responsible for representing function, performance requirements, and operational infrastructure. The functional requirements model defines what services the network must provide to the user and what services are available. The security model defines access control requirements on various system resources and specifies constraints

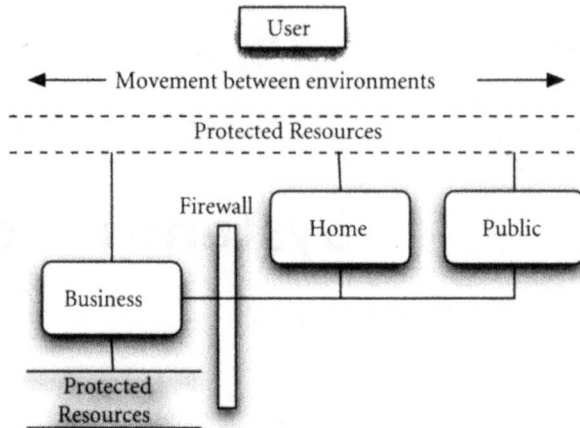

Figure 20.1 System-level networking requirements.

on how functions are provided. Finally, the infrastructure models define support provided by the various network infrastructures. If the infrastructure can provide required services while meeting system requirements, then we have an ideal system. However, modeling what the system can provide is as important as providing a binary correctness result.

Access requirements can be viewed as invariants on system state. Specifically, no network entity that is not trusted should be allowed to access private resources. Because requirements define invariants, the static domain is the most appropriate domain for an abstract specification. Functionality and infrastructure models differ from access requirements in that a need exists to specify temporal requirements. When a request is processed by the infrastructure, it must respond to that request by changing what it knows about its resources. Likewise, a requesting entity changes state when its requests are met or denied.

Because there are no timing requirements specified, the most appropriate domain here is the state_based domain, where change can be represented, but the details of that change can be avoided. The finite_state and infinite_state domains may also be appropriate, but nothing is known about the cardinality of the state set. Therefore, infrastructure and functional requirements models are written in the state_based domain.

20.1.1 Types, Constants, and Predicates

The network_entities package defined in Figure 20.2 defines a collection of shared types, functions, and constants that are shared across specifications. Virtually any large specification will define a package like network_entities that provides shared constructs to all models. A similar package can be found in the RTL case study in Chapter 17.

```
package network_entities()::static is
  id :: type is constant;
  resource :: type is constant;

  secure_server :: resource is constant;
  public_server :: resource is constant;
  internal :: id is constant;
  external :: id is constant;
  public :: id is constant;
  private :: id is constant;
  firewall :: id is constant;

  message :: type is data
    request(a,b::id;r::resource)::request?
    | grant(a,b::id;r::resource)::grant?
    | revoke(a,b::id;r::resource)::revoke?
    | trust(a,b::id)::trust?
    | distrust(a,b::id)::distrust?
  end data;

  trusts_with(a,b::id;r::resource)::boolean;

  trusts(a,b::id)::boolean;

  grants(a,b::id; r::resource)::boolean;

  transitive_trust(a,b,c::id)::boolean is
    trusts(a,b) and trusts(b,c) => trusts(a,c);

  transitive_trusts_with(a,b::id; r::resource)::boolean is
    trusts(a,b) and trusts_with(b,c,r) => trusts_with(a,c,r)

  will_grant(a,b::id; r::resource)::boolean is
    trusts_with(a,b,r) and requests(a,b,r) => grants(a,b,r)

end package network_entities;
```

Figure 20.2 Basic resource and credential entities.

The id and resource types define uninterpreted types for entities in the network and resources they may request. These types are uninterpreted because their internal representation is not germane to what is being specified. It is sufficient to define properties for entities and resources in the specification.

Several constants are defined of type resource and id that are visible to all specifications using the package. The two server declarations represent resources that may be requested and are controlled by the secure infrastructure. The two id declarations define entities that are known in all specifications. Again, the details of entities and resources are not important, thus they are declared as constants without specific values.

The message data type defines messages that are exchanged between network entities. It is critical that these definitions exist in a shared package, because all component specifications must be able to send and receive messages for requests and updates to trust relationships. Specific message instances exist for requesting, granting, and revoking access as well as for establishing and removing trust relationships.

The final element of the `network_entities` package defines a collection of predicates that define properties of components that grant access to resources. The declarations of `trusts_with`, `trusts`, and `grants` define relations between entities and resources: `trusts_with` is true if one entity trusts another with specific resource; `trusts` is true if one entity trusts the assertions of another.

The `transitive_trusts`, `transitive_trust`, and `will_grant` predicates define how trust is established and the conditions for granting access to a resource. First, `transitive_trust` simply defines and names a transitive property for trust relationships; `transitive_trusts_with` defines a relation such that if a trusts b and b trusts c with r, then a trusts c with r. Finally, `will_grant` defines a property asserting when one entity will grant access to a resource.

The way these predicates establish trust and grant resource access is relatively simple. If a resource r is requested from m by n, and m trusts n with that resource, access is granted. Establishing that m trusts n with the resource may result from first-hand knowledge. However, if m trusts an entity p that trusts n with r, then m also trusts n with r. Trust is transitive, allowing the formation of chains from an initial trusted source. This is an exceptionally naive trust model, but will suffice for this case study.

Note that none of the defined properties is asserted in the package. The properties give names to relationships, but do not assert those relationships for any particular entities. For the properties to hold, they must be asserted as terms in models that use the `network_entities` package.

20.1.2 Access Control Requirements Model

The access control requirements model defines a collection of security constraints that the network infrastructure must satisfy while providing services. It defines system-level constraints on the function of every infrastructure element in the network. While the functional requirements define what the network must do, access control requirements define what the network must not do. In this sense, access control requirements define safety conditions.

Figure 20.3 defines the access control requirements model. A component representation is chosen over a facet representation to allow inclusion of usage assumptions and correctness conditions. Usage assumptions explicitly document assumptions made when defining requirements. In effect, they define conditions that the user must accept or satisfy to use this system. Implications define correctness conditions over definitions. Thus, they support specification of theorems that should hold as a result of assumptions and definitions. If the theorems do not hold, then something about the specified requirements is incorrect.

The only assumption made is that every entity trusts `trusted_id`. This trusted entity forms the root of all trust relationships in this system. It also serves as a

```
use network_entities();
component networkSecurity()::static is
begin
  assumptions
    forall(a::id | trusts(a,trusted_id));
  end assumptions;
  definitions
  end definitions;
  implications
    // if a grants b r, then a must trust b with r
    (forall a::id |
      (forall b::id |
        (forall r::resource | grants(a,b,r) => trusts_with(a,b,r))));
  end implications;
end component networkSecurity;
```

Figure 20.3 Basic access control requirements model.

potential single point of failure. If `trusted_id` is compromised, then the entire network may fail. Making the assumption that all entities trust `trusted_id` prevents this case from occurring in this analysis. In essence, the assumption is that `trusted_id` will never be untrusted.

The requirements model is quite simple and asserts only that if a resource is granted to an entity, then the granter must continue to trust the grantee with the resource as long as the grant is valid. If `grants(a,b,r)` holds, then `trusts_with(a,b,r)` must also hold or requirements are violated. If `grants(a,b,r)` does not hold, then the implication is immediately true. The `static` domain is used because there is no need to represent change, as requirements will be defined as an invariant that must hold in all states. Note that if `trusts_with(a,b,r)` holds, it is not necessary for a to grant r to b.

Note that the requirements model uses an implication to assert its correctness condition, rather than making it a requirement. This is a stylistic choice concerning the way errors will discovered. Making the condition an implication asserts that the condition must be provable from requirements. Thus, if requirements violate the condition, the inconsistency will be detected when the condition is checked. If the condition were asserted as a requirement, the inconsistency exists in the specification immediately and can be detected without checking the implication. Unfortunately, this implies that the inconsistency could be used to verify other conditions if it is not recognized. For this reason, an implication is used rather than a definition term.

20.1.3 Functional Requirements Model

Modeling constraints outside the context of required functionality is not useful. For example, it is quite possible to define a secure network by simply not providing connectivity. If no resources are accessed and no resources are available, the network meets ideal security requirements. Of course, this is not desirable — a network without connectivity does not meet even its basic functional requirements. Thus the functional requirements model defines a minimal set of system requirements, that must be achieved in the content of constraints.

The functional requirements model defines desired functionality from the perspective of the user. It is parameterized over the identity of the sender, allowing the analysis of requirements from the perspective of a sender, behind the firewall or outside the firewall. Specifically, the send_as design parameter can be internal, external, or another id that indicates where the requester is. Thus, correctness analysis can be performed from several perspectives. Figure 20.4 defines resources that the portable computer user would ideally like to have available.

```
use network_entities();
component networkFunction
   (send_as::design id;
     i::input message;
     o::output message)::state_based is
  eventually(f::boolean)::boolean is
    not(f) implies eventually(f');
begin
  assumptions
  end assumptions;
  definitions
    eventually(o=request(send_as,private,private_server));
    eventually(o=request(send_as,private,public_server));
  end definitions;
  implications
    case send_as of
      {internal} -> eventually(i=grant(private))
                    and eventually(i=grant(public))
      {external} -> eventually(i=grant(public))
    end case;
  end implications;
end facet networkFunction;
```

Figure 20.4 Basic functional requirements.

20.1.4 Infrastructure Models

Having declarations of entities and resources, and a requirements model defining system requirements, network infrastructure models must be defined. Specifically, an internal infrastructure representing the work environment, an external infrastructure representing the home and public infrastructure, and a firewall environment represent the firewall behavior will now be defined.

Each infrastructure model defines various implementations of functionality that must satisfy system-level constraints. Thus, infrastructure models must satisfy access control requirements while defining their own requirements. This will be modeled by defining trust properties for each network environment and then conjuncting system-level requirements with each environment model.

The private infrastructure defined in Figure 20.5 represents the network environment within the organization that controls computing resources private_server, and public_server, defined earlier. It represents the world behind the organization's firewall. All infrastructure models use the state_based domain because trust relationships must change. The static domain is more appropriate for requirements because they cannot change. In effect, they define invariants on the state of network infrastructure elements.

The private infrastructure makes three trust assumptions: (i) it trusts the firewall and it trusts who the firewall trusts with the private server, (ii) it trusts

```
use network_entities();
component networkPrivateInfrastructure
  (i::input message; o::output message;
   self::design id)::state_based is
begin
  assumptions
    trusted_firewall: trusts(self,firewall)
        and forall(x:id | trusts(firewall,x) =>
                        trusts_with(firewall,x,private_server));
    trusted_internal: trusts_with(self,internal,private_server)
                      and trusts_with(self,internal,public_server);
    trusted_external: trusts_with(self,external,public_server);
  end assumptions;
  definitions
    // incoming request
    request_incoming: request?(i) and a(i)=self =>
        if trusts_with(self,b(i),r(i))
          then o'=grant(self,b(i),r(i))
                and grants(self,b(i),r(i))'
          else o'=revoke(self,b(i),r(i))
                and not(grants(self,b(i),r(i)))'
        end if;

    // incoming trust and distrust.
    trust_incoming: trust?(i) => trusts(a(i),b(i))'=true;
    distrust_incoming: distrust?(i) and trusts(a(i),b(i)) =>
        trusts(a(i),b(i))'=false;
    trust_maintenance: trusts(a,b) and not(i=distrust(a,b)) =>
        trusts(a,b)'=true;

    // maintain grants
    grant_maintenance: if grants(self,b,r)
                          and not(trusts_with(self,b,r))
        then o'=revoke(self,b,r) and not(grants(self,b,r))'
        else grants(self,b,r)'
      end if;
  end definitions;
  implications
  end implications;
end component networkPrivateInfrastructure;
```

Figure 20.5 Private network infrastructure.

internal entities with both servers, and (iii) it trusts external entities with only the public server. These three assumptions allow it to grant access to both servers to internal requesters and only the public server to external requesters. The firewall is trusted, thus the firewall can act as an arbiter for establishing trust.

The requirements model consists of five terms that represent processing incoming requests, processing incoming trust assertions and de-assertions, and maintaining trust and existing resource grants over state change. The first, `request_incoming`, processes an incoming resource request. If the node trusts the requester with the resource, a grant message is issued and the grant is maintained locally. If the node does not trust the requester, the grant is revoked and the local record of the grant is negated. It is here that we first see the use of `state_based` to move from one system state to the next, based on grants and revocations.

The `trust_incoming` and `distrust_incoming` terms are similar to the `request_incoming` term. They react to a specific input message and revise trust relations in the next state. The `trust_maintenance` term defines when trust is maintained over state change. When an entity is trusted in the current state, it is trusted in the next state unless a message has arrived that revokes that trust.

Finally, the `grant_maintenance` term maintains valid resource grants from one state to the next. It states that if the trust chain is broken and cannot be re-established, then the grant is revoked. Otherwise, the grant is kept in place in the next state. The system is thus required to recheck trust and grant relationships each time the system changes state.

EXAMPLE 20.1

Internal Request for the Private Server

As an example of how the infrastructure models operate, we can examine the scenario where an internal agent requests the private server. To initiate the request, the following message arrives at the private infrastructure's input:

```
request(private,internal,private_server)
```

Assuming that `self=private`, the message will cause the `request_incoming` term to be instantiated as follows:

```
request_incoming: true and self=self =>
      if trusts_with(self,internal,private_server)
         then o'=grant(self,internal,private_server)
              and grants(self,internal,private_server)'
         else o'=revoke(self,internal,private_server)
              and not(grants(self,internal,private_server))'
      end if;
```

The implication condition is true and the assumption `trusted_internal` makes the **if** condition true. Therefore the private infrastructure sends the message:

```
grant(private,internal,private_server)
```

and asserts:

```
grants(self,internal,private_server)
```

in the next state.

■

EXAMPLE **20.2**

External Request for the
Private Server

As an example of how the infrastructure models operate, we can examine the scenario where an external agent requests the private server. To initiate the request, the following message arrives at the private infrastructure's input:

```
request(private,external,private_server)
```

Assuming that self=private, the message will cause the request_incoming term to be instantiated as follows:

```
request_incoming: true and self=self =>
      if trusts_with(self,external,private_server)
         then o'=grant(self,external,private_server)
            and grants(self,external,private_server)'
         else o'=revoke(self,external,private_server)
            and not(grants(self,external,private_server))'
      end if;
```

The implication condition is true. However, the **if** condition cannot be verified — it is neither true nor false. Therefore, no message is sent and nothing is asserted in the next state. The request is ignored. If the internal infrastructure knew it did not trust the external requester, then a revoke message would have been sent. ∎

EXAMPLE **20.3**

Establishing Trust with an
External Requester

As a final example of the infrastructure at work, assume that an external request for the private server is made and the firewall trusts the external requester. This scenario is modeled by the following two messages:

```
trust(firewall,external)
request(private,external,private_server)
```

The first message is processed by process_trust and establishes that firewall trusts external. Combined with the assumption that private trusts firewall, it can be established that private trusts external. This would represent a situation where a VPN link is established between the firewall and the external requester.

Now the request message is processed as in the previous example. The distinction is that now it can be established that the external requester can be trusted with the private server: private trusts firewall and trusts anyone firewall trusts with private_server. The **if** condition is thus satisfied and the grant is processed as it was in the first example. ∎

The firewall component in Figure 20.6 is nearly identical to the networkPrivateInfrastructure model, with the exception that it assumes initial trust only in the private infrastructure. It does not trust the external infrastructure in any way initially. Incoming requests are handled in exactly the same manner, except that initially the firewall controls no resources. Thus, requests for resources to the firewall will go unanswered. The firewall is perfectly capable of establishing trust with other agents — its primary function.

```
use network_entities();
component firewall
  (i::input message; o::output message;
   self::design id)::state_based is
begin
  assumptions
    trusted_firewall: trusts(self,private);
  end assumptions;
  requirements
    // incoming request
    request_incoming: request?(i) and a(i)=self =>
       if trusts_with(self,b(i),r(i))
          then o'=grant(self,b(i),r(i))
               and grants(self,b(i),r(i))'
          else o'=revoke(self,b(i),r(i))
               and not(grants(self,b(i),r(i)))'
       end if;

    // incoming trust and distrust. Can assert trust between
    // any nodes
    trust_incoming: trust?(i) => trusts(a(i),b(i))'=true;
    distrust_incoming: distrust?(i) and trusts(a(i),b(i)) =>
       trusts(a(i),b(i))'=false;
    trust_maintenance: trusts(a,b) and not(i=distrust(a,b)) =>
       trusts(a,b)'=true;

    // maintain grants
    grant_maintenance: if grants(self,b,r)
                          and not(trusts_with(self,b,r))
       then o'=revoke(self,b,r) and not(grants(self,b,r))'
       else grants(self,b,r)'
       end if;
  end requirements;
  implications
  end implications;
end component firewall;
```

Figure 20.6 Firewall infrastructure.

The public infrastructure defined in Figure 20.7 differs from the private infrastructure in that it trusts everyone, but it has no resources to grant. Thus, the trust relationships are not particularly useful.

20.2 Composing System Models

Figure 20.8 defines the composition of the functional requirements model, the access control requirements model, and either the public or private infrastructure models. The parameter `public` selects the infrastructure model. Here inInfrastructure(**true**) composes the network functionality and security constraints with the public infrastructure model. In contrast, inInfrastructure(**false**)

```
use network_entities();
component networkPublicInfrastructure
  (i::input message; o::output message;
   self::design id)::state_based is
begin
  assumptions
    trusts_all: forall(x::id | trusts(self,x));
  end assumptions;
  definitions
    // incoming request
    request_incoming: request?(i) and a(i)=self =>
       if trusts_with(self,b(i),r(i))
          then oᵀ=grant(self,b(i),r(i))
               and grants(self,b(i),r(i))'
          else o'=revoke(self,b(i),r(i))
               and not(grants(self,b(i),r(i)))'
       end if;

    // incoming trust and distrust.
    trust_incoming: trust?(i) => trusts(a(i),b(i))'=true;
    distrust_incoming: distrust?(i) and trusts(a(i),b(i)) =>
       trusts(a(i),b(i))'=false;
    trust_maintenance: trusts(a,b) and not(i=distrust(a,b)) =>
       trusts(a,b)'=true;

    // maintain grants
    grant_maintenance: if grants(self,b,r)
                          and not(trusts_with(self,b,r))
       then o'=revoke(self,b,r) and not(grants(self,b,r))'
       else grants(self,b,r)'
       end if;
  end definitions;
  implications
  end implications;
end component networkPublicInfrastructure;
```

Figure 20.7 Public network infrastructure.

```
use network_entities();
use static_state_based();
facet inInfrastructure(public::design boolean) :: state_based is
  m::message;
begin
  infra: if public=true
     then networkFunction(external,m,m)
     else networkFunction(internal,m,m)
  end if;
  public: networkPublicInfrastructure(m,m,public);
  firewall: firewall(m,m,firewall);
  private: networkPrivateInfrastructure(m,m,private)
     * static_state_based.gamma(networkSecurity());
end facet inInfrastructure;
```

Figure 20.8 Composite system model.

composes the network functionality and security constraints with the private infrastructure model. Recall that the requirements model defines different requirements based on the location of the requesting entity.

The default interaction between state_based and static domains moves the requirements specification to the state_based domain. It is then composed with the private infrastructure model to define constraints on how resources are accessed. In effect, the requirements model defines a state invariant over all states. These requirements behave like assertions, monitoring the behavior of the infrastructure.

What this specification does not concern itself with is just as important — potentially more important than what is specified. No attempt is made to coordinate messages on m, as this is outside the scope of the analysis being performed. No attempt is made to model timing issues or to model how authentication is performed. This model deals quite simply with the establishment and maintenance of trust relationships and their effects on resource access. No attempt is made to deal with implementation issues. As such, this is purely a requirements model.

20.3 Constructing the Analysis Models

The inInfrastructure model defines a configurable model that defines and composes functional models, constraints, and implementation infrastructure. Analysis of these models can be achieved using a number of techniques by moving the inInfrastructure model to a domain that supports analysis. The state_based model can be moved to a simulation domain, much like previous examples. However, this would require refinement to the discrete_time domain. More appropriate here is analysis involving model checking or theorem proving, due to the logical nature of the specifications.

Constructing the analysis model uses the same type of functor as for the power-aware example. Here, a state_based model is moved to a model-checking domain for analysis:

```
state_based_mc.gamma(inInfrastructure(true));
```

The resulting model can be analyzed using model-checking techniques to assure that the invariant is met in each state and to ensure that functional requirements involving sequences of states are satisfied.

20.4 Wrap Up

This case study represents a significant departure from digital system design and looks exclusively at requirements analysis. Previous studies examine the impacts of local design decisions on system-level behavior. Here, exploring system requirements without consideration for implementation technologies is the objective.

Regardless, the impacts of local design decisions are explored in the context of system requirements.

The local design decisions explored in this case study are requirements placed on different network environments and components. Access control requirements can be varied, trust models can be altered or changed, and even models for establishing trust in the system can be changed. Here the local decisions are not implementation related but are instead requirements related.

What this case study demonstrates is a capability for modeling requirements sets before an implementation direction is chosen. System design folklore suggests that the cost of fixing an error increases by an order of magnitude for each design cycle entered prior to discovery of the error. The only thing questionable about this approximation is whether the cost should increase by only a *single* order of magnitude. Thus, the value of exploring requirements to detect errors and optimize requirements prior to system implementation cannot be overstated.

Bibliography

[Acc02] Accellera. *SystemVerilog 3.0: Accellera's Extensions to Verilog*, 2002.

[AK01] Perry Alexander and Cindy Kong. Rosetta: Semantic support for model-centered systems-level design. *IEEE Computer*, 34(11):64–70, November 2001.

[Ash96] P. Ashenden. *The Designers Guide to VHDL*. Morgan Kaufmann Publishers, San Mateo, CA, 1996.

[BCSS98] Per Bjesse, Koen Claessen, Mary Sheeran, and Satnam Singh. Lava: Hardware design in haskell. In *Proceedings of the Third ACM SIGPLAN International Conference on Functional Programming*, pages 174–184, Baltimore, MD, September 1998. In *ACM SIGPLAN Notices*, 34(1), January 1999.

[BPA95] P. Baraona, J. Penix, and P. Alexander. Vspec: A declarative requirements specification language for vhdl. In Jean-Michel Berge, Oz Levia, and Jacques Rouillard, editors, *High-Level System Modeling: Specification Languages*, volume 3 of *Current Issues in Electronic Modeling*, chapter 3, pages 51–75. Kluwer Academic Publishers, Boston, MA, 1995.

[BPSV01] J. R. Burch, R. Passerone, and A. L. Sangiovanni-Vincentelli. Overcoming heterophobia: Modeling concurrency in heterogeneous systems. In *Proceedings of the Second International Conference on Application of Concurrency to System Design*, Newcastle upon Tyne, UK, June 2001.

[CDE+] M. Clavel, F. Duran, S. Eker, P. Lincoln, N. Marti-Oliet, J. Meseguer, and J. Quesada. *Maude: Specification and Programming in Rewriting Logic*. SRI International. System Documentation.

[dAH01] Luca de Alfaro and Thomas A. Henzinger. Interface automata. In *ESEC/FSE 01: Proceedings of the Joint 8th European Engineering Conference and 9th Symposium on the Foundations of Software Engineering*, pages 109–120, Vienna, Austria, September 2001.

[Dav00] J. Davis et al. Ptolemy II—Heterogeneous concurrent modeling and design in java, University of California, Berkeley, CA, 2000.

[Dij92] Edsger J. Dijkstra. On the economy of doing mathematics. In R. S. Bird, C. C. Morgan, and J. C. P. Woodcock, editors, *Mathematics of Program Construction*, volume LNCS 669, pages 2–10. Springer-Verlag, New York, NY, 1992.

[EM85] H. Ehrig and B. Mahr. *Fundamentals of Algebraic Specifications 1: Equations and Initial Semantics*. EATCS Mongraphs on Theoretical Computer Science. Springer-Verlag, Berlin, 1985.

[FKN+92] A. Finkelstein, J. Kramer, B. Nuseibeh, L. Finkelstein, and M. Goedicke. Viewpoints: A framework for integrating multiple perspectives in system development. *International Journal of Software Engineering and Knowledge Engineering*, 2(1):31–58, March 1992. World Scientific Publishing Co.

[GH93] John V. Guttag and James J. Horning. *Larch: Languages and Tools for Formal Specification*. Springer-Verlag, New York, NY, 1993.

[GLMS02] T. Grötker, S. Liao, G. Martin, and S. Swan. *System Design with SystemC*. Kluwer Academic Press, Dordrecht, The Netherlands, 2002.

[Gri81] D. Gries. *The Science of Programming*. Texts and Monographs in Computer Science. Springer-Verlag, New York, NY, 1981.

[Gro97a] The UML Group. *UML Metamodel*. Rational Software Corporation, Santa Clara, CA, 1.1 edition, September 1997. http://www.rational.com.

[Gro97b] The UML Group. *UML Semantics*. Rational Software Corporation, Santa Clara, CA, 1.1 edition, July 1997. http://www.rational.com.

[GWM⁺93] Joseph Goguen, Timothy Winkler, José Meseguer, Kokichi Futatsugi, and Jean-Pierre Jouannaud. Introducing OBJ. In Joseph Goguen, editor, *Applications of Algebraic Specification Using OBJ*. Cambridge, MA, 1993.

[Heh93] Eric C. R. Hehner. *A Practical Theory of Programming*. Texts and Monographs in Computer Science. Springer-Verlag, New York, NY, 1993.

[Hen96] Thomas A. Henzinger. The theory of hybrid automata. In *Proceedings of the 11th Annual IEEE Symposium on Logic in Computer Science*, pages 278–292, New Brunswick, NJ, July 1996.

[HHK01] Thomas A. Henzinger, Benjamin Horowitz, and Christoph Meyer Kirsch. Giotto: A time-triggered language for embedded programming. *Lecture Notes in Computer Science*, 2211:166+, 2001.

[Hoa85] C.A.R. Hoare. *Communicating Sequential Processes*. Prentice-Hall, Englewood Cliffs, NJ, 1985.

[Hud00] P. Hudak. *The Haskell School of Expression*. Cambridge University Press, New York, NY, 2000.

[Hut99] Graham Hutton. A tutorial on the universality and expressiveness of fold. *Journal of Functional Programming*, 9(4):355–372, 1999.

[::I] Iso 10646-1:2001 unicode 3.2 standard. ISO Standard.

[IEE94] *IEEE Standard VHDL Language Reference Manual*. IEEE, New York, NY, 1994.

[IEE95] *Standard Verilog Hardware Description Language Reference Manual*, IEEE, New York, NY, 1995.

[Jon90] C. B. Jones. *Systematic Software Development Using VDM*. International Series in Computer Sience. Prentice Hall, New York, NY, 2ⁿᵈ edition, 1990.

[Jon03] Simon Peyton Jones. *Haskell 98 Language and Libraries*. Cambridge University Press, New York, NY, 2003.

[JR97] Bart Jacobs and Jan Rutten. A tutorial on (co)algebras and (co)induction. *EATCS Bulletin* 62: 222–259, 1997.

[KA02a] C. Kong and P. Alexander. Modeling model of computation ontologies in Rosetta. In *Proceedings of the Formal Specification of Computer-Based Systems Workshop (FSCBS'02)*, Lund, Sweden, March 2002.

[KA02b] C. Kong and P. Alexander. Multi-faceted requirements modeling and analysis. In *Proceedings of the IEEE Joint International Requirements Engineering Conference (RE'02)*, Essen, Germany, September 9–13, 2002.

[Mee95] Lambert Meertens. Category theory for program construction by calculation. CWI, Amsterdam and Department of Computing Science, Utrecht University, The Netherlands, September 1995.

[MFP91] Erik Meijer, Maarten Fokkinga, and Ross Paterson. Functional programming with bananas, lenses, envelopes and barbed wire. In J. Hughes, editor, *Proceedings of the 5th ACM Conference on Functional Programming Languages and Computer Architecture, FPCA'91, Cambridge, MA, USA, 26–30 Aug 1991*, volume 523, pages 124–144. Springer-Verlag, Berlin, 1991.

[Mil99] Robin Milner. *Communicating and Mobile Systems: The π-Calculus*. Cambridge University Press, New York, NY, 1999.

[Mog90] Eugenio Moggi. An abstract view of programming languages. Technical Report ECS-LFCS-90-113, Dept. of Computer Science, Edinburgh University, Edinburgh, Scotland, 1990.

[Mog91] Eugenio Moggi. Notions of computation and monads. *Information and Computation*, 93(1):55–92, 1991.

[ORS92] S. Owre, J. Rushby, and N. Shankar. PVS: A prototype verification system. In D. Kapur, editor, *Proceedings of 11th International Conference on Automated Deduction*, volume 607 of *Lecture Notes in Artificial Intelligence*, pages 748–752, Saratoga, NY, June 1992. Springer-Verlag, Berlin.

[Pie91] B. Pierce. *Basic Category Theory for Computer Scientists*. The MIT Press, Cambridge, MA, 1991.

[Pie02] Benjamin C. Pierce. *Types and Programming Languages*. The MIT Press, Cambridge, MA, 2002.

[Pie03] Benjamin C. Pierce. Types and programming languages: The next generation. Slides presented at *IEEE Symposium on Logic in Computer Science (LICS)*, Ottawa, Canada, June 2003.

[Sch86] David A. Schmidt. *Denotational Semantics: A Methodology for Language Development*. Allyn & Bacon, Inc., Boston, MA, 1986.

[Smi90] Douglas R. Smith. KIDS: A Semiautomatic Program Development System. *IEEE Transactions on Software Engineering*, 16(9):1024–1043, 1990.

[Spi92] J. M. Spivey. *The Z Notation: A Reference Manual.* International Series in Computer Science. Prentice Hall, New York, NY, 2nd edition, 1992.

[Sto77] Joseph E. Stoy. *Denotational Semantics: The Scott-Strachey Approach to Programming Language Theory.* The MIT Press, Cambridge, MA, 1977.

[TM96] D. E. Thomas and P. R. Moorby. *The Verilog Hardware Description Language.* Kluwer Academic Publishers, Boston, MA, 3rd edition, 1996.

[UV05] Tarmo Uustalu and Varmo Vene. The essence of dataflow programming. In K. Yi, editor, *Proceedings of APLAS'05—Lecture Notes in Computer Science*, volume 3780, pages 2–18. Springer-Verlag, Berlin, 2005.

[VHL89] I. Van Horebeek and J. Lewi. *Algebraic Specifications in Software Engineering: An Introduction.* Springer-Verlag, Berlin, 1989.

[Wad92] P. Wadler. The essence of functional programming. In *Conference Record of the Nineteenth Annual ACM SIGPLAN-SIGACT Symposium on Principles of Programming Languages*, pages 1–14, Albuquerque, New Mexico, January 1992.

[Wad97] P. Wadler. How to declare an imperative. *ACM Computing Surveys*, 29(3):240–263, 1997.

[WD96] Jim Woodcock and Jim Davies. *Using Z: Specification, refinement and Proof.* Prentice Hall, Upper Saddle River, NJ, 1996.

[Win90] J. Wing. A Specifier's Introduction to Formal Methods. *IEEE Computer*, 23(9):8–24, September 1990.

[Wor96] J. B. Wordsworth. *Software Engineering with B.* Addison Wesley Longman Ltd., Harlow, England, 1996.

Index